Lectures on Mathematics in the Life Sciences
Volume 2

SOME MATHEMATICAL QUESTIONS IN BIOLOGY

The American Mathematical Society
Providence, Rhode Island
1970

Proceedings of the Second Symposium on Mathematical Biology held in New York, December, 1967; and the Third Symposium on Mathematical Biology held in Dallas, December, 1968. Prepared by the American Mathematical Society under Contract with the National Institutes of Health, Contracts PH43-68-43(1967) and NIH-69-749(1968).

International Standard Book Number 0-8218-1152-5
AMS 1970 Subject Classification 92A05

Printed in the United States of America

CONTENTS

FOREWORD

There is still much debate over whether mathematical biology is an existing science or merely a hope for the future. If it is a science per se, then the classic works of Lotka[1] in 1924 and Volterra[2] in 1931 are amongst the early fundamental contributions. The age of these works approaches a half century, a period of time which, nowadays, has been adequate for a physical theory to reach maturity. Nevertheless, we can at best say only that it is a nascent science, the major directions of which are at most imperfectly known. Since biological systems superficially seem to elude physical laws, or at least to depend on ones not yet known (cf. Schrödinger[3]), we have been wary of applying the limited mathematical and physical theories which *are* available to us. Biological systems are enormously complex. Thus, when applying our limited mathematical tools we tend to be tempted by the beauty and simplicity of the mathematics and accept a mathematically elegant description of something unreal rather than an unsatisfying, even

[1] A. J. Lotka, *Elements of mathematical biology*, Dover, New York, 1956 (Originally published as *Elements of physical biology*, Williams and Wilkins, 1924).

[2] V. Volterra, *Leçons sur la théorie mathématique de la lutte pour la vie*, Gauthier-Villars, Paris, 1931.

[3] E. Schrödinger, *What is life?*, Cambridge Univ. Press, Cambridge, 1951, pp. 68–69.

unworkably complex, description of what is actually there. Hence, the development of mathematical biology has been understandably slowed.

There is an additional obstacle which can be illustrated by viewing another science which has been slow to come of age—mathematical linguistics. Language is as old as civilization, and the study of language as old as trade. Yet the modern science of language is as young as the combinations of electronic-computer-plus-software, in which we have seen the embodiment of grammars of certain man-made languages. There are as yet no artificial systems in mathematical biology by which we can increase our understanding or test our methods, although the creation of synthetic genes seems to give promise that some of these may be available before long. For the present, we can only abstract from the tremendous complexity of nature, with obvious accompanying limitations.

Until the question of whether mathematical biology is a science or a hope is resolved, we say only that these are "Symposia on Some Mathematical Questions in Biology". The papers of Mac Arthur and Cowan were presented at the second of these symposia, held in New York, on December 27, 1967; those of Levins and Winfree were given at the third, held in Dallas, on December 28, 1968. The symposia are jointly sponsored by the American Mathematical Society and the Society for Industrial and Applied Mathematics, and are held annually in conjunction with the American Association for the Advancement of Science. We wish finally to acknowledge with gratitude that the symposia of which this volume forms the proceedings, were supported by grants from the National Institutes of Health.

Murray Gerstenhaber
Department of Mathematics
UNIVERSITY OF PENNSYLVANIA
Philadelphia, Pa. 19104
June 12, 1970

A STATISTICAL MECHANICS OF NERVOUS ACTIVITY

By

JACK D. COWAN*

The University of Chicago

To Warren McCulloch and Walter Pitts

In memoriam

* Supported in part by the Office of Naval Research (Physics Branch).

Contents

Introduction

One of the most striking features of the electrical activity of the brain is that save for catastrophic states such as death or coma, it is both incessant and ubiquitous. Anyone who has ever looked at electroencephalograms will confirm this. Which are the neural structures that generate such activity? Exactly how is the activity generated? What is the functional significance of the activity? In what way is the activity related to the firing patterns of individual nerve cells in the brain?

This paper contains a first attempt at a quantitative study of the activity generated by interconnected aggregates of model nerve cells. It is intended that statistical properties of this activity be compared with those obtained from actual recordings of nervous tissues. Therefore the method of statistical mechanics has been used to study the model aggregates. The basic cell properties are outlined and an equation representing cell responses is derived in § 1. Interconnected aggregates are then considered in § 2, and a Hamiltonian mechanics is derived of the resulting activity. This leads naturally to a Gibbs' ensemble theory of the variations in the activity. Such a theory is considered in § 3 where the level of neuronal activity (measured by cell firing rates) is shown to be the formal analogue of energy, and the amplitude of fluctuation in the activity the formal analogue of temperature. In § 4 applications are made of the theory to hypothetical cortico-thalamic nets that supposedly generate maintained neural activity. A number of statistical features of the activity are derived and a preliminary comparison is made with experimental data.

1. The neural equation

1a. **Cell properties.** The theory outlined in this paper is not concerned with the spatio-temporal aspects of nervous activity, but only with the temporal aspects. The actual spatial relations of interconnected cells are ignored, only topological features being of interest, so that Boolean matrices suffice to specify cellular interconnections. Each cell consists of a soma and dendrites which collect incoming excitation, and an axon which transmits the cell's own excitation to other cells. It is assumed that when any (primary) cell becomes active it instantaneously transmits an impulse of excitation to all the (secondary) cells to which it is connected.

MEMBRANE POTENTIAL. Let E_r be the resting potential of the cell membrane, and v the transmembrane potential or simply the membrane potential relative to the cell resting potential. Let $h_{ij}(t - t')$ be the deviation in the membrane potential of the ith secondary cell at time t caused by an impulse transmitted from the jth primary cell at time t'. It is assumed that $h(t)$ is the function

$$h_{ij}(t) = u_{-1}(t)\, \delta v_{ij} \exp\,[-t/\tau_m]$$

where $u_{-1}(t)$ is the Heaviside step-function (equal to zero for $t < 0$, and to one for $t \geq 0$), and where δv_{ij} is the deviation in the membrane potential caused by the arrival of an impulse from the jth cell. In the appendix it is shown that

$$\delta v_{ij} = (\sigma/C_m)(\delta g_{ij} E_j)$$

where σ is the effective operating period for the transmission of excitation from one cell to another, E_j the change in the membrane potential caused by the action of an impulse from the jth cell, and δg_{ij} the corresponding change in membrane conductance. An equivalent form is also derived:

$$\delta v_{ij} = \delta Q_{ij}/C_m$$

where δQ_{ij} is the charge delivered to the ith cell membrane by the action of an impulse from the jth cell, and where C_m is the capacitance of the cell membrane. It is also shown that the relaxation time-constant of the membrane potential is $\tau_m = C_m/g$ where g is the total membrane conductance.

Cells are assumed to be either excitatory or else inhibitory. For an excitatory cell the change in the membrane potential of any secondary cell to which it transmits excitation is proportional to $-E_m$, for an inhibitory cell the change is proportional to $E_i - E_m$. It is assumed that $E_i < E_m < 0$, so that $-E_m$ is positive and $E_i - E_m$ is negative.

The cells are assumed to sum the incoming excitation in linear fashion. In the appendix it is shown that under such an assumption the membrane potential built-up in the ith secondary cell by the action of N primary cells is represented by the convolution

$$v_i(t) = \sum_{j=1}^{N} \int_0^t h_{ij}(t - \tau) f_j(\tau)\, d\tau$$

where $f_j(\tau)$ is the mean frequency of arrival of impulses from the jth primary cell. It is assumed that these frequencies are slowly varying compared with $1/\tau_m$ the reciprocal of the membrane time constant. As shown in the appendix the membrane potential then takes the simple form

(1) $$v_i(t) = (\sigma/g_m)(\sum_j \delta g_{ij} E_j f_j(t))(1 - \exp(-t/\tau_m))$$

where $\tau'_m = C\tau_m$, and $\sigma = a\,\delta t/b$ as defined in the appendix.

THRESHOLD. When the membrane potential reaches a certain threshold value θ, changes occur in the membrane conductance g_m that give rise to a further rapid increase and subsequent relaxation of the membrane potential known as the action-potential, i.e., the cell becomes active and transmits an impulse of excitation to other cells. As is well known the Hodgkin-Huxley equations adequately represent the sequence of changes in g_m and the resulting action-potential, at least for axons (Hodgkin and Huxley [**12**]). However the equations are equivalent to a fourth-order nonlinear differential equation with time-varying coefficients which has not been solved in closed form. It is therefore difficult to obtain analytical results on the responses of the Hodgkin-Huxley equations for "noisy" stimuli (cf: Stein [**26**]). Since the object of this paper is to model nets rather than single cell activities, a detailed representation of single cell responses is not attempted. Instead the mean firing rate of a cell is assumed to be related to the mean value of the current built-up in the cell membrane by incoming excitation. In particular $f_i(t)$ the mean rate at which the ith cell emits impulses is taken to be the function:

(2) $$f(t) = [r(1 + \exp(-\beta[(i(t)/i_{th}) - 1]))]^{-1}$$

where

$$\beta = \left(1 - \frac{i_0}{i_{th}}\right)^{-1} \ln\left(\frac{1 - rf_0}{rf_0}\right).$$

The constant r is the refractory period of the neuron, that period following the emission of an impulse when the cell is absolutely inexcitable; the threshold current i_{th} is that current which drives the cell at the 50% rate; i_0 is approximately $1.5i_{rh}$ where i_{rh} is the "rheobase", the smallest current which will elicit an impulse from the cell and v_0 the corresponding firing rate.

Two remarks are called for. Firstly, there is no fundamental reason for choosing such a function, at least none that can be seen at present. The chosen function is of course the well-known logistic curve of demography (cf: Pearl [**22**], Lotka [**20**]). It happens to be a convenient and tabulated function that fits quite well a certain amount of data on the mean responses of cells to stimulating currents (Creutzfeldt [**6**], Robertson [**12**]). What is of interest is the qualitative nature of the results that follow such a choice. There is at present no good reason for choosing any one function over another except that the logistic function is one of the simpler sigmoidal curves to work with. This is in keeping with the main object of this study which is entirely heuristic. I am attempting to devise a model neural net which is both mathematically tractable yet not completely devoid of biological relevance in order to learn how to think about neural nets and how to design experiments. The second remark I want to make is that it will not have escaped the knowledgeable reader that the present formulation applies only to "tonic" cells which give maintained responses to constant stimuli, and not to "phasic" cells which respond only to changing stimuli. In a later paper I hope to show how such phasic cells can be modeled.

1b. **Net properties.** As I have already noted only topological features of nets are of interest, spatial relationships are to be ignored. However there are certain general properties of the nets I want to examine that follow immediately from two assumptions on cell properties. These assumptions are that cells are either excitatory or else inhibitory at all their synaptic terminals (cf: Eccles [**8**]), and that excitation is transmitted instantaneously from cell to cell. The consequences of these assumptions are the following. Consider any random neural net. It consists of random distributions of both excitatory and inhibitory cells, randomly interconnected. Suppose the cells are ordered in some arbitrary fashion, then there exists a random matrix representing the connectivity of the net. Merely by reordering, the net can be partitioned into two nets; one consisting only of excitatory cells, the other of inhibitory cells, and the connectivity will be represented by a random matrix in which any row consists entirely of nonnegative or else nonpositive elements. Figure 1 shows the resulting "canonical" form for neural nets that follows from the assumptions.

1c. **The neural equation.** To derive an equation relating the neural firing rates $f_i(t)$ and $f_j(t)$ is quite simple. Equation (1) gives for the mean current built-up in the cell membrane, the formula:

$$\langle i_i(t) \rangle = \left(\frac{c\sigma}{C_m}\right) \sum_j \delta g_{ij} E_j f_j \langle \exp(-t/c\tau_m) \rangle$$

where $\langle \quad \rangle$ denotes some suitable time-average taken such that the rates $f_j(t)$ are quasi-static. The time-average $\langle \exp(-t/c\tau_m) \rangle$ is simply $\alpha_i/c\tau_m$ where $0 < \alpha_i < 1$, so

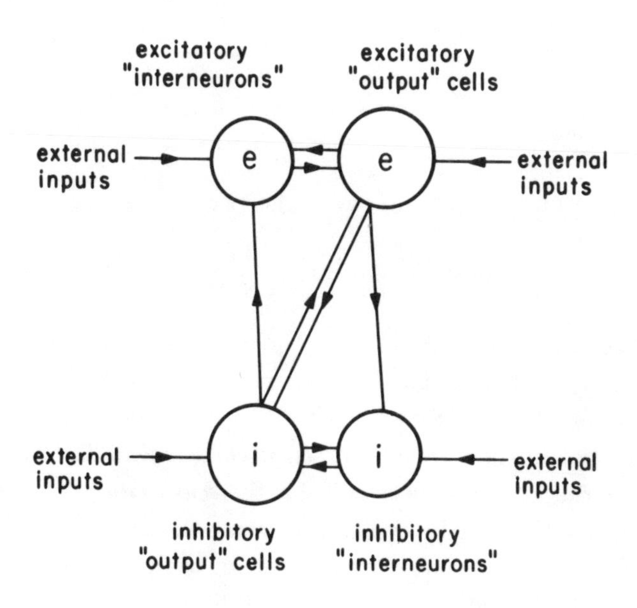

FIGURE 1
The "canonical" form for a neural net.

that

$$\langle i_i(t) \rangle = \frac{c\sigma\alpha_i}{C_m} \sum_j \delta g_{ij} E_j f_j(t).$$

Equation (2) therefore becomes

$$f_i(t) = \left[r\left(1 + \exp\left(-\frac{\beta c \sigma \alpha_i}{C_m i_{th}}\right) \sum_j \delta g_{ij} E_j f_j(t) + \beta\right)\right]^{-1}$$

Let

$$x(t) = 1 - rf(t) \tag{3}$$

and

$$\delta g_{ij} E_j = \alpha_{ij},$$

$$\frac{\beta c \sigma \alpha_i}{C_m i_{th} r} = \frac{1}{\beta_i},$$

$$\varepsilon_i = \beta - \frac{1}{\beta_i} \sum_j \alpha_{ij},$$

then Equation (2) may be further written as:

$$\ln \frac{x_i(t)}{1 - x_i(t)} = \varepsilon_i + \frac{1}{\beta_i} \sum_j \alpha_{ij} x_j(t). \tag{4}$$

The variable x may be interpreted as the fraction of time during a sufficiently long interval when a cell that is continually emitting spikes is not refractory, i.e., the fraction of time when it is "sensitive" to incoming stimuli. For this reason I have called x the *sensitivity* of the cell. Clearly the variable $1 - x$ represents the fraction of time when the cell is refractory and cannot be fired by incoming stimuli, i.e., the insensitivity. For sufficiently large samples we can use a central result of renewal theory (Cox, [**5**]) that the renewal density, here the mean interspike interval τ, is inversely proportional to the mean rate f, so that one can also write

$$x = 1 - r/\tau \tag{5}$$

thus relating sensitivity to interspike intervals.

The sensitivity function is in fact loosely related to certain functions occurring in the renewal theory of Geiger counters. A type 1 counter is one which is blocked or refractory following the reception of a particle, for a certain duration. The neuron model I am using may be thought of as the combination of a type 1 counter with a coincidence detector. Let r be the refractory or blocked time of the system and f the rate of arrival of θ particles, where θ is the threshold of the detector. It is well known [**5**] that the mean failure-time during which the system is insensitive is $r + 1/f$ and that N_t, the number of coincidences counted in an interval of duration t, is asymptotically normally distributed with mean $ft(1 + rf)^{-1}$. A large sample estimate for f can be obtained by equating this with N_t, so that $ft/(1 + rf) = N_t$ or $f = (N_t/t)/[1 - r(N_t/t)]$. If $f_i(t)$ is identified with N_t/t, the counting-rate of the system, one sees that the estimate of the rate of arrival of sufficiently many inputs to fire the model neuron is given by the quantity $1/r \cdot [(1 - x)/x]$. This provides a heuristic justification for using the sensitivity function, although it is certainly not a rigorous way of connecting it with renewal theory.

It is now necessary to derive from Equation (4) which is valid for quasi-stationary inputs, an equation that is valid for nonstationary inputs, i.e., a dynamical equation for neural responses. To do this properly is a very difficult proposition (Johannesma [**13**]); I am therefore going to make yet another heuristic assumption, that the next step beyond the quasi-stationary behavior represented by Equation (4) is given by:

$$\left(\tau \frac{d}{dt} + 1\right) \ln \left(\frac{x_i}{1 - x_i}\right) = \varepsilon_i + \frac{1}{\beta_i} \sum_j \alpha_{ij} x_j. \tag{6}$$

This then is the neural equation, an ordinary nonlinear differential equation in the variable x. In the next section I shall outline some of its properties. I wish only to remark here that Equation (6) represents a model neuron that is in a sense a generalization of the McCulloch-Pitts neuron [**21**]. The McCulloch-Pitts neuron has a threshold nonlinearity which is discontinuous, and is concerned with the emission of individual impulses. The model neuron I have introduced above has a threshold nonlinearity which is continuous (but in the limit it can approximate the McCulloch-Pitts threshold nonlinearity), and is concerned not with individual impulses, but with the continuous variable x related to the probability of impulse emission.

2. Interconnected aggregates

2a. **The general neural net.** The general equation for any net can now be written, given the assumptions I have made. Let $x\rangle$ be the column vector representing the "state" of the net, i.e., the pattern of cell sensitivities at time t, and let $u\rangle$ be a column vector representing the external "inputs" to each cell in the net. Let $f(x)$ be the function $\ln(x/1 - x)$. Then the general equation for any net is

$$\left(\tau \frac{d}{dt} + 1\right) f(x\rangle) = Bu\rangle + Ax\rangle$$

where B is a constant matrix, and A the matrix of coupling coefficients, an $n \times n$ square matrix.

It follows from § 1 that this equation can always be rewritten as the two equations

$$\begin{aligned} \left(\tau \frac{d}{dt} + 1\right) f(x_1\rangle) &= B_1 u_1\rangle + A_{11} x_1\rangle - A_{12} x_2\rangle \\ \left(\tau \frac{d}{dt} + 1\right) f(x_2\rangle) &= B_2 u_2\rangle + A_{21} x_1\rangle - A_{22} x_2\rangle \end{aligned} \tag{7}$$

where the indices 1 and 2 refer respectively to the aggregates of inhibitory and of excitatory cells, where B_1 and B_2

are arbitrary "control" matrices, and where the matrices A_{ij}, $i, j = 1, 2$ are random matrices with nonnegative elements.

What insights can be obtained from these equations about the steady-state and transient activities of neural nets? Given a large-scale net of the type I have introduced, it is obvious, I think, that one cannot hope to analyze in detail any of the exact features of the activity. The exact solution for the activity of a single cell will not furnish any useful information about the large-scale activity. The reason of course is that the individual cell rapidly loses its identity as the size of the net increases. Indeed in a net with very many cells it is no longer possible to specify the state of the net at any time t, simply because the net is neither completely observable nor completely controllable. Even in the unlikely event that a solution existed for the activity of the net, it would be unusable because of the indeterminacy for all practical purposes of the initial state. The conclusion is obvious: one must look at *time-averages* of the activity, not at the activity itself. This in turn implies the use of *statistical mechanics* in its most general sense as seen in the investigations of Prigogine [**31**], Kubo [**18**], Zwanzig [**30**], Rice and Gray [**24**] and others, on the nonequilibrium statistical mechanics of many-body systems. In the present paper I am not going to attempt such a development, but I will attempt to sketch only the equilibrium statistical mechanics (Gibbs [**11**]) of special neural nets and the closely related Gaussian-noise approach to such nets.

2b. **An analogy with population dynamics.** As it happens, most of the basic machinery for this has already been developed by Kerner [**14**] and [**15**] in his investigations of the ecological kinetics first proposed by Volterra [**27**]. Furthermore Leigh [**19**] has amplified some of Kerner's results and himself applied the Gaussian-noise approach to further elucidate the Volterra-Kerner kinetics. In their most general form these equations can be written as

$$\frac{d}{dt} \ln x\rangle = \varepsilon\rangle + Ax\rangle$$

where $x\rangle$ is a state vector the ith coordinate of which represents the equivalent number of individuals belonging to the ith species at time t, where $\varepsilon\rangle$ is a vector the ith coordinate of which represents the rate at which the ith species grows or decays, and where A is the interaction matrix. The so-called "predator-prey" interaction is of particular relevance to neural net problems: here the association consists of species that are either predatory or else are eaten by predators. In this case the equations become:

$$\begin{aligned} \frac{d}{dt} \ln x_1\rangle &= \varepsilon_1\rangle - A_{11}x_1\rangle - A_{12}x_2\rangle \\ \frac{d}{dt} \ln x_2\rangle &= \varepsilon_2\rangle + A_{21}x_1\rangle - A_{22}x_2\rangle \end{aligned} \tag{8}$$

It will be seen that these equations are quite close to Equation (7) for neural nets. This suggests that if excitatory cells correspond with predators and inhibitory cells with prey, the analogy will be most apposite. Of course there are important differences in the structure of these equations, in particular the structure of the damping terms in the neural equations is more complex. For the moment however I am going to investigate the properties of a special system of equations that corresponds most closely with the Volterra-Kerner predator-prey system, namely the equations

$$
\tau \frac{d}{dt} \ln \frac{x_1\rangle}{1 - x_1\rangle} = B_1 u_1\rangle - A_{12} x_2\rangle, \tag{9}
$$
$$
\tau \frac{d}{dt} \ln \frac{x_2\rangle}{1 - x_2\rangle} = B_2 u_2\rangle + A_{21} x_1\rangle.
$$

To make it easier to see what is going on, I will write these

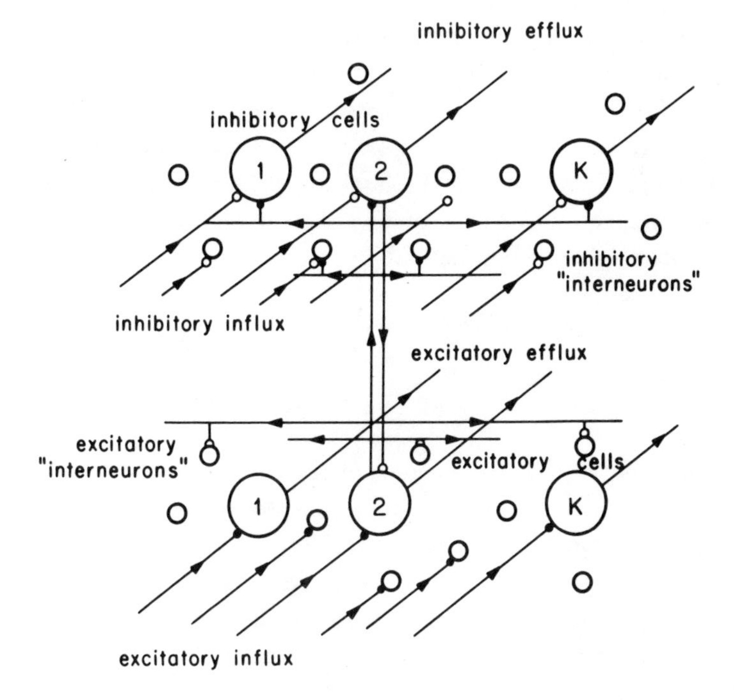

FIGURE 2
Interactions of excitatory and inhibitory cells corresponding to Equation (10).

equations in the following form:

$$(10)\qquad \tau\frac{dx_i}{dt} = \left(b_iu_i + \frac{1}{\beta_i}\sum_j^n \alpha_{ij}x_j\right)x_i(1-x_i)$$

with the subsidiary conditions

$$(11)\qquad \alpha_{ij} + \alpha_{ji} = 0, \qquad \alpha_{ii} = 0.$$

Thus I am looking at the very special net shown in Figure 2, in which there is no self-excitation or inhibition, in which the cells are all tonic, and in which damping is assumed to be negligible for the time being.

2c. **A neural "Hamiltonian".** Equation (10) is now in a form very closely related to the Volterra-Kerner equations, and what follows merely parallels Kerner's application of Gibbsian statistical mechanics.

Let me first transform Equation (10) in the following fashion. It is clear that the stationary states of the net are either x_i equal to zero or one, or else they are given by the solution of the equation

$$\beta_ib_iu_i + \sum_j^n \alpha_{ij}q_j = 0,$$

that is

$$(12)\qquad q_j = \sum_i^n A'_{ji}\beta_ib_iu_i$$

where A'_{ji} is the jith element of the inverse of A_{ij}. Such an inverse exists and is unique whenever A_{ij} is skew-symmetric and of even order. If $\beta_ib_iu_i$ lies within the range $(-\sum_j \alpha_{ij}, 0)$ for excitatory cells, and within $(0, -\sum_j \alpha_{ji})$ for inhibitory cells, then these stationary states lie within the interval (0,1). Now consider the following transformation:

$$(13)\qquad v_i = \ln\frac{(x_i/q_i)}{(1-x_i)}, \qquad T = t/\tau.$$

Then Equation (10) may be rewritten as

$$\frac{dv_i}{dT} = \sum_j^n \left(\frac{\alpha_{ij}}{\beta_i \beta_j}\right) \frac{\partial G}{\partial v_j}, \tag{14}$$

where

$$G = \sum_i \beta_i (\ln (1 + q_i \exp v_i) - q_i v_i). \tag{15}$$

G is of course the "Hamiltonian" for the net, the first integral of the system. It is easy to show that it is a positive-definite function, and that the equations represent a conservative dynamical system that exhibits undamped oscillations in the variables x_i or v_i about the stationary states q_i. So for this special case I have obtained an equation of motion for the activities of a finite net of model neurons, that is essentially in Hamiltonian form and in fact corresponds to an aggregate of coupled non-linear oscillators, the oscillations being in neural sensitivities or effectively in interspike intervals.

In passing it is of interest that Wiener and Pitts in the late 1940s appear to have been working along somewhat similar lines, but apart from the following remarks, nothing was published:

> The anatomical picture of the cortex suggests that we may usefully employ statistical methods in the study of its function. This work has been taken up brilliantly by Walter Pitts. He finds that under many conditions, the approximation to the activity of the cortex should not be made on a basis of rest and the zero basal activity. Proceeding from this he has developed the wave-equations of the cortex. These

promise to give an interpretation of the electroencephalogram. He has also suggested the exploration of the connections between different parts of the cortex by suitable stimulation of regions and the observance of the frequency response at several points.

N. Wiener [**29**]

3. Neural Gibbs ensembles

3a. **Neural phase space, Liouville's equation, and a canonical distribution for neural sensitivities.** The machinery has now been set up for a straightforward application of Gibbsian statistical mechanics. The steps are very well documented, and explicit constructions are given by Kerner and by Leigh [**19**] of the standard ensembles. I will not repeat these steps here. It suffices to say that a "phase-space" is considered, each point of which has coordinates $v_1, v_2, \ldots, v_{2n}$ (this phase-space is also a "state-space" or "configuration-space", since the generalized coordinates v_i are not distinguishable as coordinates and momenta). It is easily shown that any sufficiently large collection of points in the space—the *ensemble*—acts like an incompressible fluid so that any function of the form $\int\int \cdots \int D \, dv_1 \, dv_2 \cdots dv_{2n}$, where D is an arbitrary function of the $2n$ integrals of Equation (10), is an invariant of the motion in the phase space induced by the equations. If D is taken to be the relative density of members of the ensemble within the volume element $dv_1 \, dv_2 \cdots dv_{2n}$ in phase-space, then an invariant probability measure is obtained, suitable for computing ensemble-averages. The final step is to take the relative density to be a function of the *first* integral of the equations i.e., of the Hamiltonian G, and to choose a particular form for this function. This is usually done in one or two ways; either by choosing a density the form of which is additive for noninteracting subsystems, or else by choosing a density that optimizes some aspect of the system as measured by an observer, or else optimizes the control problem for the system. In all these cases the result is the same: given a large isolated system with Hamiltonian G a monotone increasing additive function of the system's components, then any small subsystem comprising say $2k$ of the $2n$ components of the system will have a relative density given by

$$\varphi = a \exp\left(-b \sum_{1}^{2k} G_i\right) \tag{16}$$

where a is a normalizing constant, and where b will be defined later. This density is of course the familiar "canonical" density of statistical physics.

In what follows this density will be used to compute ensemble averages of various quantities associated with the special neural net I have constructed. The net is assumed to be a small subset of a very large system in a state of equilibrium (i.e., the inputs $bu\rangle$ are to be held constant), and the rest of the system acts like a kind of "heat-bath" for the net.

3b. **Calculation of ensemble averages.** I will now use the standard machinery I have set up to compute ensemble averages taken with respect to the canonical density. It is easily shown that

$$\overline{\partial G / \partial v_i} = 0.$$

But

$$\frac{\partial G}{\partial v_i} = \beta q_i \left(\frac{\exp v_i}{1 + q_i \exp v_i} - 1\right) = \beta_i(x_i - q_i).$$

Therefore the ensemble average of x_i equals the stationary state q_i, $\overline{x_i} = q_i$.

Let me now compute directly the time-average of x_i:

$$\langle x_i \rangle = \lim_{T \to \infty} \frac{1}{T} \int_0^T x_i(\alpha)\, d\alpha.$$

It follows from Equation (10) that

$$\frac{\beta_i}{T} \ln \frac{y_i(T)}{y_i(0)} = \beta_i b_i u_i + \sum_j \alpha_{ij} \left(\frac{1}{T} \int_0^T x_j(\alpha)\, d\alpha \right)$$

where $y_i = (x_i/1 - x_i)$. Therefore $\langle x_i \rangle = q_i$. Combining these two results one sees that:

$$\overline{x_i} = \langle x_i \rangle.$$

Thus the time-average of x_i equals the canonical ensemble average. This is a very useful property, and suggests that the property of ergodicity can be invoked for the system, so that *all* ensemble averages may be replaced by time-averages. It is known that the great majority of non-integrable Hamiltonian systems possess the so-called quasi-ergodic property (Birkhoff [**2**]). I shall assume in what follows that all time-averages commute with ensemble averages. Let me first give an interpretation for the constant b. It is easily shown that

$$\overline{v_i \frac{\partial G}{\partial v_i}} = \frac{1}{b} \tag{17}$$

and that

$$\overline{\beta_i (x_i - q_i)^2} = \overline{\left(\frac{\partial G}{\partial v_i} \right)^2}$$

$$= \frac{\left(\frac{1}{b}\right) \beta_i q_i (1 - q_i)}{1 + \left(\frac{1}{b}\right) \beta_i}. \tag{18}$$

The quantity $1/b$ is essentially the analog of the "kinetic energy" of the net, and so the interesting result is forthcoming that this energy is *equipartitioned* throughout the net. This suggests that the constant $1/b$ may be taken to be the formal analog of temperature for the net. This can be seen more clearly if equation 18 is inverted, so that:

$$\frac{1}{b} = \theta = \beta_i \frac{\overline{(x_i - q_i)^2}}{q_i(1 - q_i)} \Big/ \left(1 - \frac{\overline{(x_i - q_i)^2}}{q_i(1 - q_i)}\right). \tag{19}$$

It will be seen that θ is closely related to the mean-square deviation of any variable x_i about its corresponding mean value q_i. For this reason, following Kerner, I will call the Hamiltonian G the *activity* of the net, and θ the *amplitude of fluctuation* of this activity.

Having provided an interpretation for θ, let me now find a formula for the constant a, and hence for the canonical density. By definition,

$$\begin{aligned} a &= \int_{-\infty}^{\infty}\int_{-\infty}^{\infty} \cdots \int_{-\infty}^{\infty} \exp\,(-G/\theta)\, dv_1\, dv_2 \cdots dv_{2k} \\ &= \prod_{1}^{2k} \int_{-\infty}^{\infty} \exp\,(-G_i/\theta)\, dv_i \\ &= \prod_{1}^{2k} \exp\,(-\beta_i/\theta) \int_{-\infty}^{\infty} [\ln\,(1 + q_i \exp v_i) - q_i v_i]\, dv_i \qquad (20) \\ &= \prod_{1}^{2k} (q_i^{-\beta_i q_i/\theta}) \int_{0}^{1} x_i^{(\beta_i q_i/\theta)-1}\,(1 - x_i)^{[\beta_i(1-q_i)/\theta - 1]}\, dx_i \\ &= \prod_{1}^{2k} (q_i^{-\beta_i q_i/\theta}) B(\beta_i q_i/\theta,\, \beta_i(1 - q_i)/\theta) \end{aligned}$$

where $B(p,q)$ is Euler's β-function.

Given this formula it is now an easy step to write down the canonical density, that is the probability that any one cell will have a sensitivity x_i within the range x_i, $x_i + dx_i$.

This is simply

$$(21)\quad \begin{aligned} p(x_i)\,dx_i &= \frac{q_i^{-\beta_i q_i/\theta} x_i^{\beta_i q_i/\theta - 1}(1 - x_i)^{\beta_i(1-q_i)/\theta - 1}\,dx_i}{q_i^{-\beta_i q_i/\theta} B(\beta_i q_i/\theta,\, \beta_i(1 - q_i)/\theta)} \\ &= \frac{x_i^{\beta_i q_i/\theta - 1}(1 - x_i)^{\beta_i(1-q_i)/\theta - 1}\,dx_i}{B(\beta_i q_i/\theta,\, \beta_i(1 - q_i)/\theta)}\,. \end{aligned}$$

This density is immediately recognizable as the well-known *β-density* of mathematical statistics (Feller [**10**]) and population genetics (Kimura [**17**]), usually written in the form:

$$(22)\qquad p(x)\,dx = B(p,q)^{-1}\,x^{p-1}(1 - x)^{q-1}\,dx \qquad (p,q > 0).$$

The range of the density is from $x = 0$ to $x = 1$. Its distribution function is the incomplete β-function, normalized by the β-function. When both p and q exceed unity, the density is zero at its extremities and has a unique mode at $(p - 1)/(p + q - 2)$. In the cases of relevance to neural net problems, $p = \beta_i q_i/\theta$ and $q = \beta_i(1 - q_i)/\theta$ so that $p + q = \beta_i/\theta$. Two conditions, the unimodality of the density, and the finiteness of θ serve to determine the maximum value of $\overline{x_i^2}$ and hence of the variance of the fluctuations in x_i. Using such conditions one obtains the inequality:

$$\overline{(x_i - q_i)^2} < \frac{\min\,(q_i,\, 1 - q_i)}{1 + \min\,(q_i,\, 1 - q_i)}\, q_i(1 - q_i).$$

The right-hand side of this equation is zero at its extremities $q_i = 0$ and 1, and takes its maximum value of $\frac{1}{12}$ at $q_i = \frac{1}{2}$. One can see from this that the peak to peak amplitude of variation of any neural oscillator is a maximum whenever the stationary state equals $\frac{1}{2}$, and tends to zero as the stationary state approaches 0 or 1.

The moments of the β-density are

$$\int_0^1 x^k p(x)\, dx = \frac{\Gamma(p+k)\Gamma(p+q)}{\Gamma(p)\Gamma(p+q-1)}$$

so that the mean is

$$\mu = P/(p+q) \tag{23}$$

and the variance

$$\sigma^2 = (pq)/(p+q)^2(p+q+1). \tag{24}$$

From Equation (23) one sees that $\bar{x}_i = q_i$ as before.

If one compares the mean with the mode of the density it is seen that

$$q_i \gtreqless \frac{\dfrac{\beta_i q_i}{\theta} - 1}{\dfrac{\beta_i}{\theta} - 2} \quad \text{whenever} \quad q_i \lesseqgtr 1 - q_i.$$

Thus when $q_i < \frac{1}{2}$ the distribution weights values of x_i are less than q_i. When $q_i > \frac{1}{2}$ the distribution weights values of x_i are greater than q_i, and when $q_i = \frac{1}{2}$ the mean and the mode coincide. Figure 3 depicts some examples of the oscillations about various values of q_i. It will be seen that the distribution correctly predicts the general nature of the oscillations. In fact these can be investigated much more closely by examining the fractions T_+/T and T_-/T of a long time-interval T spent by an oscillator above and below the mean level q_i. By ergodicity these fractions are simply the distribution functions $Pr(x_i > q_i)$ and $Pr(x_i < q_i)$ respectively. Thus

$$T_-/T = Pr(x_i < q_i) = I_{q_i}(p,q) \tag{25}$$

and

$$
\begin{aligned}
T_+/T &= Pr(x_i > q_i) \\
&= 1 - Pr(x_i < q_i) \qquad (26) \\
&= I_{1-q_i}(q,p)
\end{aligned}
$$

where $I_\alpha(p,q)$ is the incomplete β-function. Once again the way in which the distribution weights the activity may be seen directly from these formulas.

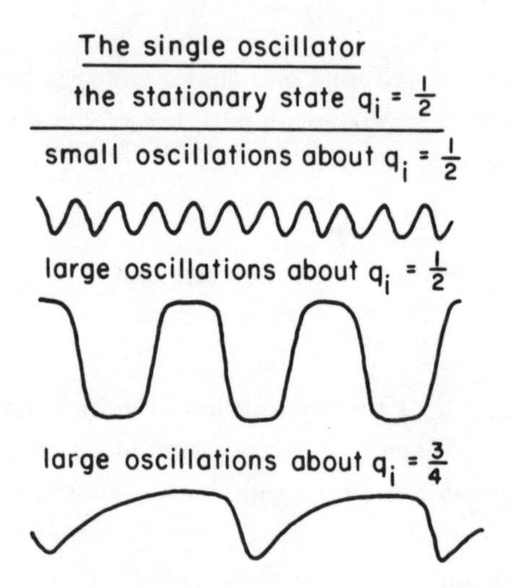

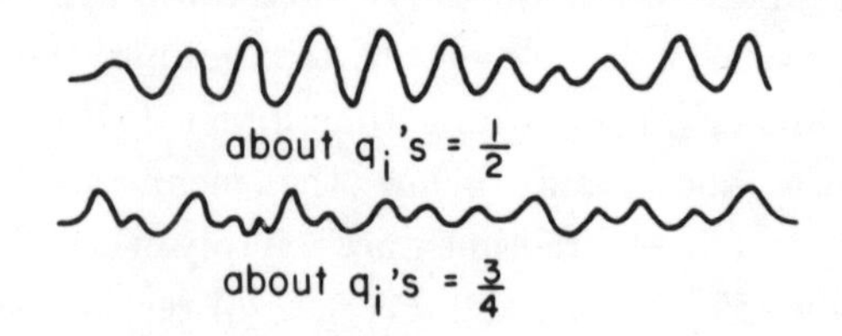

FIGURE 3
Oscillations in "sensitivity."

I now want to turn to the calculation of two very important "observables" of neural activity, the mean rate at which neural fluctuations occur, and cross-correlation functions of the joint activity of two or more cells. Let me first consider the crossing-rate of the variable x_i about some fixed value x. The time average of this rate is given by $\langle|\dot{x}_i|\,\delta(x_i - x)\rangle$ which by ergodicity equals the ensemble average

$$\overline{|\dot{x}_i|\,\delta(x_i - x)}.$$

This can be developed as follows:

$$\begin{aligned}&\overline{|\dot{x}_i|\,\delta(x_i - x)}\\ &= \int_0^1\int_0^1\cdots\int_0^1 |\dot{x}_i|\,\delta(x_i - x)p(x_1\,x_2\cdots x_{2k})\,dx_1\,dx_2\cdots dx_{2k}.\end{aligned}$$

But $x_i \in (0,1)$ so that $x_i(1 - x_i) > 0$, hence

$$\begin{aligned}|\dot{x}_i| &= \left|\left(b_i u_i + \frac{1}{\beta_i}\sum_j \alpha_{ij}x_j\right)x_i(1 - x_i)\right|\\ &= x_i(1 - x_i)\left|b_i u_i + \frac{1}{\beta_i}\sum_j \alpha_{ij}x_j\right|,\end{aligned}$$

and recalling that

$$\begin{aligned}p(x_1\,x_2\cdots x_{2k}) &= \prod_1^{2k} p(x_i),\\ p(x_i) &= \frac{x_i^{\lambda_i-1}(1 - x_i)^{\mu_i-1}}{B(\lambda_i,\mu_i)},\\ \overline{|\dot{x}_i|\,\delta(x_i - x)} &= B(\lambda_i,\mu_i)^{-1}\int_0^1 x_i^{\lambda_i}(1 - x_i)^{\mu_i}\,\delta(x_i - x)\,dx_i\\ &\quad\times\int_0^1\int_0^1\cdots\int_0^1\left|b_i u_i + \frac{1}{\beta_i}\sum_j \alpha_{ij}x_j\right|\\ &\quad\times\prod_1^k p(x_j)\,dx_j.\end{aligned}$$

(I have introduced the notation $\lambda_i = \beta_i q_i/\theta$, $\mu_i = \beta_i(1 - q_i)/\theta$ in order to avoid confusion.)

The first of these integrals is easily evaluated, and gives a contribution

$$x_i^{\lambda_i}(1 - x_i)^{\mu_i}/B(\lambda_i, \mu_i)$$

towards the product.

The second multiple integral can be evaluated in the following fashion. Recall that $\beta_i b_i u_i + \sum_j \alpha_{ij} q_j = 0$. Therefore

$$\beta_i b_i u_i + \sum_j \alpha_{ij} x_j = \sum_j \alpha_{ij}(x_j - q_j).$$

Furthermore let $h_+(x_j - q_j)$ and $h_-(x_j - q_j)$ be functions such that

$$\begin{aligned} h_+(x_j - q_j) &= 1, x_j > q_j, \\ &= 0, x_j < q_j, \\ h_-(x_j - q_j) &= 1 - h_+(x_j - q_j). \end{aligned} \tag{27}$$

It is then easily seen that

$$\begin{aligned} \overline{|x|} &= \overline{(h_+(x) - h_-(x))x} \\ &= 2\overline{h_+(x)x} - \overline{x} = -2\overline{h_-(x)x} + \overline{x} \end{aligned}$$

so that

$$\overline{\left|\sum_j \left(\frac{\alpha_{ij}}{\beta_j}\right)(x_j - q_j)\right|} = -2\overline{h_-\left(\sum_j \left(\frac{\alpha_{ij}}{\beta_j}\right)(x_j - q_j)\right) \times \left(\sum_j \left(\frac{\alpha_{ij}}{\beta_j}\right)(x_j - q_j)\right)}$$

since

$$\overline{\sum_j \left(\frac{\alpha_{ij}}{p_j}\right)(x_j - q_j)} = 0.$$

Thus the second integral becomes

$$-2\int_0^1\int_0^1 \cdots \int_0^1 h_-\left(\sum_j \left(\frac{\alpha_{ij}}{\beta_j}\right)(x_j - q_j)\right) \times \left(\sum_j \left(\frac{\alpha_{ij}}{\beta_j}\right)(x_j - q_j)\right) \prod_1^k p(x_j)\, dx_j.$$

Let me take the simplest case first in which $k = 1$, so that the net consists of one excitatory cell interacting with one inhibitory cell. The integral reduces to

$$-2\int_0^1 h_+\left(\left(\frac{\alpha_{ij}}{\beta_j}\right)(x_j - q_j)\right)\left(\frac{\alpha_{ij}}{\beta_j}\right)(x_j - q_j)\,dx_j$$

$$= -2\left(\frac{\alpha_{ij}}{\beta_j}\right)\int_0^{q_j}(x_j - q_i)p(x_j)\,dx_j$$

$$= -2\left(\frac{\alpha_{ij}}{\beta_j}\right)B(\lambda_j,\mu_j)^{-1}(B_{q_j}(\lambda_j + 1, \mu_j) - q_j B_{q_j}(\lambda_j,\mu_j))$$

where

$$B_{q_j}(\lambda_i, \mu_j) = \int_0^{q_j} x_j^{\lambda_j - 1}(1 - x_j)^{\mu_j - 1}\,dx_j$$

is the incomplete β-function.

But it is easily shown that

$$B_{q_j}(\lambda_j + 1, \mu_j) = \frac{\lambda_j}{\lambda_j + \mu_j}\,B_{q_j}(\lambda_j, \mu_j) - \frac{q_j^{\lambda_j}(1 - q_j)^{\mu_j}}{\lambda_j + \mu_j}\,.$$

Using this, the integral can be reduced to the expression:

$$2\left(\frac{|\alpha_{ij}|}{\beta_i\beta_j}\right)\theta\cdot\frac{q_j^{\lambda_j}(1 - q_j)^{\mu_j}}{B(\lambda_j,q_j)}\,.$$

Thus in this simple case I can write down an explicit form for the mean crossing-rate as

$$\begin{aligned} w[x_j : x] &= \overline{|\dot{x}_i|\,\delta(x_i - x)} \\ &= 2\,\frac{|\alpha_{ij}|}{\beta_i\beta_j}\cdot\frac{x^{\lambda_i}(1 - x)^{\mu_i}}{B(\lambda_i,\mu_i)}\cdot\frac{q_j^{\lambda_j}(1 - q_j)^{\mu_j}}{B(\lambda_j, \mu_j)}\cdot\theta, \end{aligned} \tag{28}$$

or in the more transparent form,

$$\begin{aligned} w[x_j : x] = 2\,\frac{|\alpha_{ij}|}{\beta_i\beta_j}\,&x(1 - x)p(x_i = x) \\ &\times q_j(1 - q_j)p(x_j = q_j)\cdot\theta. \end{aligned} \tag{29}$$

Now let me take the case of $k = 2$, so that the net consists of two excitatory cells and two inhibitory cells. Now I need the ensemble average of the expression

$$h_{-}\left[\left(\frac{\alpha_{i1}}{\beta_i}\right)(x_1 - q_1) + \left(\frac{\alpha_{i2}}{\beta_i}\right)(x_2 - q_2)\right]$$
$$\cdot\left[\frac{|\alpha_{i1}|}{\beta_i}(x_1 - q_1) + \frac{|\alpha_{i2}|}{\beta_i}(x_2 - q_2)\right].$$

This is given by the expression

$$\frac{|\alpha_{i1}|}{\beta_i}\int_0^{\varphi_{11}}\int_0^{\varphi_{21}}(x_1 - q_1)p(x_1)p(x_2)\,dx_1\,dx_2$$
$$+\frac{|\alpha_{i2}|}{\beta_i}\int_0^{\varphi_{12}}\int_0^{\varphi_{22}}(x_2 - q_2)p(x_1)p(x_2)\,dx_1\,dx_2$$

where

$$\varphi_{11} = (|\alpha_{i1}|\,q_1 + |\alpha_{i2}|\,q_2)/|\alpha_{i1}|,$$
$$\varphi_{21} = (|\alpha_{i1}|/|\alpha_{i2}|)(\varphi_{11} - x_1),$$
$$\varphi_{22} = (|\alpha_{i1}|\,q_1 + |\alpha_{i2}|\,q_2)/|\alpha_{i2}|,$$
$$\varphi_{12} = (|\alpha_{i2}|/|\alpha_{i1}|)(\varphi_{22} - x_2).$$

That is I am integrating over the triangular region bounded by the loci $x_1 = 0$, $x_2 = 0$, and $(|\alpha_{i1}|/\beta_i)(x_1 - q_1) + (|\alpha_{i2}|/\beta_i)(x_2 - q_2) = 0$.

But

$$B(\lambda_1,q_1)\int_0^{\varphi_{11}}(x_1 - q_1)p(x_1)\,dx_1$$
$$= B_{\varphi_{11}}(\lambda_1 + 1,\, \mu_1) - q_1 B_{\varphi_{11}}(\lambda_1,\mu_1)$$
$$= \frac{-\varphi_{11}^{\lambda_1}(1 - \varphi_{11})^{\mu_1}}{(\lambda_1 + \mu_1)}$$
$$= \frac{-\theta}{\beta_1}\,\varphi_{11}^{\lambda_1}(1 - \varphi_{11})^{\mu_1}.$$

Similarly

$$B(\lambda_2,\mu_2)\int_0^{\varphi_{11}}(x_2 - q_2)p(x_2)\,dx_2 = \frac{-\theta}{\beta_2}\,\varphi_{22}^{\lambda_2}(1 - \varphi_{22})^{\mu_2},$$

and since φ_{11} and φ_{22} are independent of x_1 and x_2,

$$\int_0^{\varphi_{11}}\int_0^{\varphi_{21}}(x_1 - q_1)p(x_1)p(x_2)\,dx_1\,dx_2$$

$$= \int_0^{\varphi_{21}}(-\theta/\beta_1)\,\frac{\varphi_{11}^{\lambda_1}(1 - \varphi_{11})^{\mu_1}}{B(\lambda_1,\mu_1)}\,p(x_2)\,dx_2$$

$$= (-\theta/\beta_1)\,\frac{\varphi_{11}^{\lambda_1}(1 - \varphi_{11})^{\mu_1}B_{\varphi_{21}}(\lambda_2,\mu_2)}{B(\lambda_1,\mu_1)B(\lambda_2,\mu_2)}\,.$$

Similarly

$$\int_0^{\varphi_{12}}\int_0^{\varphi_{22}}(x_2 - q_2)p(x_1)p(x_2)\,dx_1\,dx_2$$

$$= (-\theta/\beta_2)\,\frac{\varphi_{22}^{\lambda_2}(1 - \varphi_{22})^{\mu_2}B_{\varphi_{12}}(\lambda_1,\mu_1)}{B(\lambda_2,\mu_2)B(\lambda_1,\mu_1)}\,.$$

Using these results and simplifying and collecting terms, one obtains for the mean-crossing rate the formula:

$$w[x_i:x] = x(1 - x)p(x_i = x)2\theta$$

$$\cdot\left[\frac{|\alpha_{i1}|}{\beta_i\beta_1}\,\varphi_{11}(1 - \varphi_{11})p(x_1 = \varphi_{11})I_{2,\varphi_{21}} + \frac{|\alpha_{i2}|}{\beta_i\beta_2}\,\varphi_{22}(1 - \varphi_{22})p(x_2 = \varphi_{22})I_{1,\varphi_{12}}\right] \tag{30}$$

where $I_{j,\varphi_{jj}}$ is the β-distribution function for the jth cell. The case of $k = 3$ can also be carried through quite easily to give:

$$w[x_i:x] = x(1 - x)p(x_i = x)2\theta$$

$$\cdot\left[\frac{|\alpha_{i1}|}{\beta_i\beta_1}\,\varphi_{11}(1 - \varphi_{11})p(x_1 = \varphi_{11})\int_0^{\varphi_{31}}I_{2,\varphi_{21}}p(x_3)\,dx_3 + \frac{|\alpha_{i2}|}{\beta_i\beta_2}\,\varphi_{22}(1 - \varphi_{22})p(x_2 = \varphi_{22})\int_0^{\varphi_{32}}I_{1,\varphi_{12}}p(x_3)\,dx_3 + \frac{|\alpha_{i3}|}{\beta_i\beta_3}\,\varphi_{33}(1 - \varphi_{33})p(x_3 = \varphi_{33})\int_0^{\varphi_{13}}I_{2,\varphi_{23}}p(x_1)\,dx_1\right]. \tag{31}$$

Since

$$\int_0^{\varphi_{ij}} I_{m,\varphi_{mj}} p(x_i)\, dx_i = \int_0^{\varphi_{ij}} \int_0^{\varphi_{mj}} p(x_i)p(x_m)\, dx_i\, dx_m$$
$$= \Pr\,[x_i < \varphi_{ij}, x_m < \varphi_{mj}],$$

the general expression for any k can now be written down as

$$w[x_i : x] = x(1 - x)p(x_i = x)\left(\frac{2\theta}{\beta_1}\right)$$
$$\cdot\left[\sum_j \frac{|\alpha_{ij}|}{\beta_j}\, \varphi_{ij}(1 - \varphi_{jj})p(x_j = \varphi_{jj})\right. \tag{32}$$
$$\left.\cdot \Pr\,[x_1 < \varphi_{1j}, x_2 < \varphi_{2j}, \ldots, x_t < \varphi_{tj}, t \neq j]\right]$$

where

$$\varphi_{jj} = \sum_t \frac{|\alpha_{it}|\, q_i}{|\alpha_{ij}|}, \qquad \varphi_{tj} = \frac{|\alpha_{i(t-1)}|}{|\alpha_{it}|}\,(\varphi_{i(t-1)} - x_{t-1}).$$

Following Kerner it can be seen that the mean crossing-rate about x relative to the mean crossing-rate about q_i is the simple expression:

$$w[x_i : q_i] = df\, \frac{w[x_i - x]}{w[x_i - q_i]} = \frac{x(1 - x)p(x_i = x)}{q_i(1 - q_i)p(x_i = q_i)}\,. \tag{33}$$

In principle then the canonical density can be obtained through a sequence of crossing-rate measurements on single cells in the net (as well as by the direct method of computing inter-spike interval distributions).

Alternatively this formula can be inverted to give θ, the amplitude of fluctuation, as a measurable quantity:

$$\theta = \frac{\beta_i q_i \ln \dfrac{x_i}{q_i} + \beta_i(1 - q_i) \ln \dfrac{1 - x_i}{1 - q_i}}{\ln w[x : q_i]}\,. \tag{34}$$

Thus θ or θ/β_i is obtained from q_i, x, and $w[x:q_i]$. Since this determination does not depend on k, the number of "interactions" per cell, it provides a very useful measure of the total activity of the net.

Turning now to the computation of cross-correlation functions for the joint activities of cells in a net, it is easily shown that all moments of the form $\overline{x_i^\alpha x_j^\beta}$ vanish. So the actual activities appear to be uncorrelated. However moments of the form $\overline{x_i \dot{x}_j}$ do not vanish, and provide a very useful observable. Let me introduce the notation:

$$\langle v_i(1)\dot{v}_j(1)\rangle = \frac{d}{ds}\left[\lim_{T\to\infty}\frac{1}{T}\int_{-T}^{T} v_i(t)v_j(t+s)\,dt\right]_{s=0}$$

for the cross-correlation function at one instant. By ergodicity

$$\begin{aligned}\langle v_i(1)\dot{v}_j(1)\rangle &= \overline{v_i(1)\dot{v}_j(1)} \\ &= \sum_j \left(\frac{\alpha_{k_i}}{\beta_k\beta_j}\right)\overline{v_i(1)\,\frac{\partial G}{\partial v_j(1)}}\end{aligned}$$

(by equation (14)). But

$$\overline{v_i(1)\,\frac{\partial G}{\partial v_j(1)}} \begin{array}{ll} = \theta, & i = j \\ = 0, & \text{otherwise} \end{array}$$

so that

$$\langle v_i(1)\dot{v}_k(1)\rangle = \left(\frac{\alpha_{ki}}{\beta_k\beta_i}\right)\theta. \tag{35}$$

Thus the coupling-coefficient between any two cells in the net is a function of the cross-correlation function of certain aspects of their activities, and of the amplitude of the fluctuations in their activities. Formula 35 can also be rewritten in terms of the sensitivities x_i:

$$\frac{\alpha_{ki}}{\beta_k\beta_i} = \theta^{-1}\left\langle \ln \frac{x_i(1)/q_i}{1 - x_i(1)} \frac{d}{dT} \ln \left(\frac{x_k(1)/q_k}{1 - x_k(1)} \right) \right\rangle. \tag{36}$$

I will conclude this subsection by summarizing the results I have obtained so far. For a very special net of simplified model cells I have shown that the ensemble and time averages of the quantity x_i, the sensitivity of the ith cell, are equal to the stationary-states q_i of the cell sensitivity. I have shown that there is an analogue of kinetic energy for the net, which is equipartitioned throughout all the cells in the net and therefore the analogue of temperature, the amplitude of fluctuation θ, has meaning for the net. This quantity I have related to the first two moments of the canonical density $p(x_i)\,dx_i$ of cell sensitivities, which I demonstrated to be the β-density of statistics and population genetics. I then used this density to compute such averages as the mean-crossing rates for cells in the net, and cross-correlation functions of aspects of their joint activity, and to provide a way of measuring the amplitude of fluctuation of the net's activity from single cell measurements.

3c. **A sketch of the Fokker-Planck approach to equilibrium densities.** Before turning to the all-important question of biological relevance, I want to make a few remarks about the equations I have been using, in connection with the Langevin equations used to describe the Brownian motion of small particles under various conditions (Wax [**28**]). Consider first the equation for a single cell under the influence of a stationary Gaussian white-noise source $\eta(t)$ with mean μ and variance σ^2. Let me first consider Equation (6) for a single cell with the added noise term:

$$(37) \qquad \left(\tau \frac{d}{dt} + 1\right) \ln \frac{x_i}{1 - x_i} = \eta_i(t).$$

This is easily transformed into the Langevin equation

$$(38) \qquad \tau\, dz = -z\, dt + \eta(t)$$

under the transformation

$$(39) \qquad z = \ln \frac{x_i}{1 - x_i}.$$

It is well known that corresponding to this equation there is a Fokker-Planck equation for the transition probability density $p(y, x, t)$ that $z = y$ at t' given $z = x$ at $t' - t$, namely the equation:

$$(40) \qquad \frac{\partial p}{\partial t} = -\frac{\partial}{\partial y}[a(y)p] + \frac{1}{2}\frac{\partial^2}{\partial y^2}[b(y)p]$$

where $a(y) = \mu - y/\tau$, $b(y) = \sigma^2$. The equilibrium density is then given by the solution of the equation

$$(41) \qquad \frac{\partial}{\partial y}[a(y)p] = \frac{1}{2}\frac{\partial^2}{\partial y^2}[b(y)p]$$

with suitable boundary conditions. The solution of this equation is easily obtained and is

$$p(z)\, dz = (\sqrt{\pi}\sigma)^{-1} \exp\left[-\left(\mu - \frac{z}{\tau}\right)^2 \Big/ (\sigma^2/\tau)\right],$$

so that

(42)

$$p(x_i)\, dx_i = (\sqrt{\pi}\sigma)^{-1} \exp\left[-\left(\mu - \frac{1}{\tau}\ln\frac{x_i}{1 - x_i}\right)^2 \Big/ (\sigma^2/\tau)\right] \times (x_i(1 - x_i))^{-1}\, dx_i.$$

It will be seen that this density is not transformable into a β-density. Consider however the simplified neural equation

$$\tau \frac{d}{dt} \ln \frac{x_i}{1 - x_i} = \eta(t). \tag{43}$$

This gives into the stochastic equation

$$\tau\, dz = \eta(t) \tag{44}$$

with associated Fokker-Planck equation for the equiiib-rium density, given by Equation (41) with

$$a(y) = \mu, \qquad b(y) = \sigma^2, \tag{45}$$

the solution of which is easily obtained as

$$p(z)\, dz = (c\sigma)^{-1} \exp\,(2\mu z/\sigma^2)\, dz \tag{46}$$

so that

$$\begin{aligned} p(x_i)\, dx_i &= (c\sigma)^{-1}(x_i(1 - x_i))^{-1} \\ &\qquad \times \exp\left(2\mu \ln \frac{x_i}{1 - x_i}\bigg/ \sigma^2\right) dx_i \\ &= (c\sigma)^{-1} x_i^{2\mu/\sigma^2 - 1}(1 - x_i)^{-2\mu/\sigma^2 - 1}\, dx_i. \end{aligned} \tag{47}$$

It will be seen that although this density is close to the β-density $B(p, q)^{-1} x_i^{p-1}(1 - x_i)^{q-1}$ in fact it is pathological for if $p > 0$ then $q < 0$.

Let me now look at the linearized versions of Equations (37), (39):

$$\left(\tau \frac{d}{dt} + 1\right) y_i = n_i(t) \tag{48}$$

and

$$\tau \frac{dy_i}{dt} = \eta_i(t) \tag{49}$$

where $y = 4x_i - 2$, the first approximation to $\ln (x_i/1 - x_i)$. From the Langevin theory it follows that the equilibrium densities for these cases are given respectively by

$$(50) \quad p(y_i)\,dy_i = (\sqrt{\pi\sigma})^{-1} \exp \left[-\left(\mu - \frac{y_i}{\tau}\right)^2 \Big/ (\sigma^2/\tau) \right] dy_i$$

and

$$(51) \qquad p(y_i)\,dy_i = (c\sigma)^{-1} \exp\,[2\mu y_i/\sigma^2]\,dy_i.$$

Equation (50) is of special interest. It is a well-known result that the canonical density associated with a net of harmonic oscillators (of the type we have investigated in § 3b) is also given by Equation (50). So in the case of *linear* oscillators, the equilibrium densities for a net of *undamped* oscillators, and for the activities of a collection of noninteracting *damped* elements each subject to independent random Langevin forces, are the same. In this special case then, it is possible to trade off loss of activity due to damping against the increase of activity brought about by the presence of Langevin forces. It is clear, I think, that in the nonlinear case, one *cannot* obtain an exact balance of activity in the same fashion, at least not with a simple Gaussian-noise input, and that the equilibrium densities, even for single nonlinear cells, do not correspond to the canonical densities that exist for "conservative" nets.

Similar results hold for nets under the influence of Langevin forces. I want to cite one example of the kind of result that has been obtained. Let me refer back to the vector notation I introduced in § 2a, to consider the system of linearized neural equations with Langevin terms:

$$\dot{x}\rangle = Ax\rangle + \eta\rangle. \tag{52}$$

For such a system it has been shown (Butchart [**3**]) that there is an equilibrium density of the form

$$(\pi\sigma^2)^{-k/2} \, |E^{-1}|^{-1/2} \exp\left[\langle xEx\rangle/\sigma^2\right] \tag{53}$$

where the quadratic form $\langle xEx\rangle$ is a Liapunov function of the system of equations

$$\dot{x}\rangle = Ax\rangle, \tag{54}$$

so that the matrix E is related to the matrix A, in fact $|E|$ is the product of the kth and $(k-1)$th Hurwitz determinants of A. Implicit in this result is the relationship

$$a_{ii} = -a\sigma^2, \tag{55}$$

i.e., the diagonal elements of the matrix A are proportional to the variance of the noise input $\eta(t)$. In the conservative case all the diagonal elements of A are zero, and the Liapunov function $\langle xEx\rangle$ reduces to the constant of the motion G. In such a case the equilibrium density is of course canonical, but because the diagonal elements α_{ii} equal zero, no Langevin effects can be present. If there are Langevin effects, the equilibrium density is still canonical in the Gibbs sense, but is of the form $\exp[-\beta V]$ with V a Liapunov function for the system. A reasonable conjecture is that similar results hold for the general nonlinear net that I introduced in § 2.

4. Possible applications of the formalism

Having now derived equilibrium densities for simplified neural equations (Equations (10)) and conjectured that somewhat similar results hold for more realistic equations (e.g., Equation (6)), I now wish to outline how the various ensemble averages could be used to analyze experimental data.

4a. **Thalamo-cortical circuits.** I will use Equation (10) as a model of the reciprocal interactions between neural nets in the neocortex and thalamus of the vertebrate nervous system. Figure **4** shows the scheme. Large stellate cells in a thalamic nucleus are supposed to excite the apical dendrites of pyramidal cells in the neocortex, by the "nonspecific" afferents of Ramon y Cajal [**4**]. In turn these pyramidal cells are supposed to inhibit the stellate cells by way of a cortico-thalamic projection system. In addition, both the cortical pyramids and their thalamic counterparts are embedded in a web of "interneurons". Finally there is an inhibitory input to the cortical pyramids that is assumed to be either from neighboring nets in the cortex, or from sensory inputs; and there is a corresponding excitatory input to the thalamic counterparts which is assumed to be either an "arousal" signal from subthalamic neural nets or from proprioceptive inputs. Another interpretation of these inputs is of considerable interest: the inputs, both cortical and thalamic, may be thought of as being intrinsic to the cells on which they play. The force of this interpretation is that cortical cells act *passively*, firing only when excited by thalamic cells whereas thalamic cells act as "*pacemakers*" of a kind, firing at an ever-increasing rate unless inhibited by cortical cells. On such a view the thalamus drives the cortex, which in turn modulates the thalamic activity: such an interaction evidently gives rise to oscillations of membrane potentials and currents, and of cell firing rates. Equation (10) represents a model for such a system, and so the resulting statistical mechanics can in principle provide a means for analyzing the activities of such systems.[1]

4b. **Interval histograms and amplitude densities for single cells.** Given such cortico-thalamic circuits then, many ensemble averages can be derived for the analysis of experimental data. It follows from Equations (5) and (21) that a formula for $p(\tau_i)\, d\tau_i$ the interspike interval density may be derived, namely:

$$p(\tau_i)\, d\tau_i = \frac{(1 - r/\tau_i)^{\lambda_i - 1}(r/\tau_i)^{\mu_i - 1}(r/\tau_i^2)}{B(\lambda_i, \mu_i)}\, d\tau_i\,. \tag{56}$$

It will be seen that this density is defined on (r, ∞) and is skewed with a long tail. The parameters of the density, λ_i and μ_i are given by the formulas:

$$\begin{aligned} \lambda_i &= (q_i/\theta)(C_m i_{th} r/\beta c \sigma \alpha_i), \\ \mu_i &= ((1 - q_i)/\theta)(C_m i_{th} r/\beta c \sigma \alpha_i). \end{aligned} \tag{57}$$

This density is a two-parameter one, with parameters q_i and θ. The parameter q_i is of course the stationary state of the activity of the ith cell, and is a function of both control stimuli and coupling coefficients to the cell. The equilibrium density $p(\tau_i)$ therefore reflects some of the cell's environment as well as intrinsic properties; Formulas (56) and (57) provide one explicit model for such relationships.

---

[1] The neurologically sophisticated reader will of course recognize that what I am proposing is a model for the alpha rhythms of the electroencephalograph which is not at all novel, but has been proposed many times in the last forty years (cf: Andersen and Andersson [**1**]). In my opinion the model is still the most promising one and moreover I believe that there are homologies to be found between the circuits I have proposed for cortico-thalamic interactions, and those in many other parts of the nervous system (cf: Eccles *et al.* [**9**]) so that a profitable way to look at the nervous system is to view it as composed of "universal elements" each of which comprises circuits composed of cells in the telencephalon, diencephalon, and mesencephalon. The general neural net that I discussed in § 2a would serve as a model for such vertically organized elements.

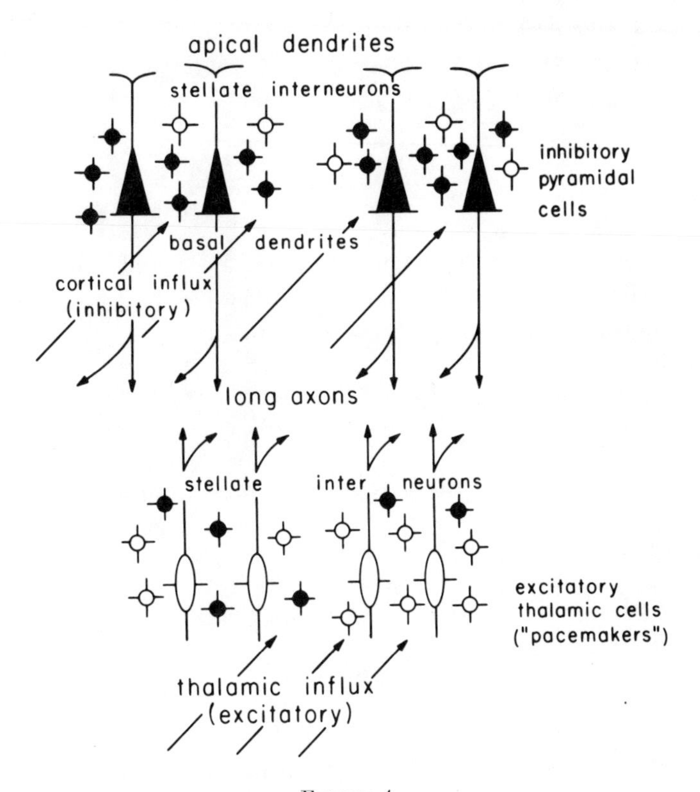

Figure 4
Thalamo-cortical circuits.

There is a difficulty concerning amplitude densities for membrane potential, in that the simplified neural equations that I have used for the Gibbs-ensemble do not give rise to a constant functional relationship between membrane potential and firing rates (or sensitivities), but rather to the relationship:

$$\ln \frac{x_i}{1 - x_i} = \beta\left(\frac{t}{\tau}\right) - \left(\frac{\alpha}{\tau}\right)v. \tag{58}$$

On the assumption that the time-constant τ is large compared with the time-constant for changing excitation, I will assume the relationship

$$\ln \frac{x_i}{1 - x_i} = -\left(\frac{\alpha}{\tau}\right) v + \left(\frac{c}{\tau}\right) \tag{59}$$

between the mean membrane potential v and the sensitivities x_i, where c is a constant. Then it is easily shown that v has an amplitude density given by:

$$\begin{aligned} p(v_i)\, dv_i = {} & B(\lambda_i, \mu_i)^{-1} \exp\left[-\tau^{-1}(\alpha v_i - c)(\beta_i q_i/\theta)\right] \\ & \times \left(1 + \exp\left[-\tau^{-1}(\alpha v_i - c)\right]\right)^{-\beta_i/\theta}(\alpha/\tau_i)\, dv_i \end{aligned} \tag{60}$$

In other words if the variable u_i is introduced such that

$$u_i = \exp\left[\tau^{-1}(\alpha v_i - c)\right], \tag{61}$$

then u_i has an amplitude density given by

$$p(u_i)\, du_i = \frac{u_i^{q-1}}{(1 + u_i)^{p+q}} B(p, q)^{-1}\, du_i. \tag{62}$$

This will be recognized as a β-density of the second kind. According to Feller [**10**] the tails of this density are of Pareto type, i.e., are transformable into a density which has no finite moments but which has the property that linear combinations of such random variables also have amplitude densities of Pareto type. This is a very important property for it permits the prediction of relationships between the firing patterns of cells and their associated membrane potentials, and the electro-encephalograms or "EEG" recorded by gross electrodes, on the obvious assumption that the EEG is to first-order a linear combination of somato-dendritic cell potentials.

4c. **The measurement of cell interactions.** A quantity of considerable interest in the analysis of neural net activities is the "coupling-coefficient" of cell interactions. Formulas (36) and (59) when combined give rise to a way of determining such coefficients, for their combination gives

$$\alpha_{ik} = + \frac{(\beta_k \alpha_k)(\beta_i \alpha_i)}{\theta \tau^2} \langle v_i(1) \dot{v}_k(1) \rangle. \tag{63}$$

Thus the coupling coefficient α_{ik} is proportional to the cross-correlation of essentially the "current" driving the kth primary cell with the "voltage" built-up in the membrane of the ith secondary cell. It will be seen that *this relation is independent of the particular nonlinearity assumed in this paper* (Equation (2)), *but depends critically on the type of dynamics assumed for cell activities.* In principle then the measurement of current-intensity relationships for individual cells (to determine β_i) and of crossing-rate measurements of the amplitude of fluctuation of the background activity (to determine θ), followed by the measurement of current-voltage correlations between cells, suffices to determine the coupling coefficients of the interaction between tonic cells.

4d. **Interval densities from the Fokker-Planck approach.** By exactly the same methods interval densities and so on may be derived from the equilibrium density obtained from the Langevin theory (Equation (42)). Thus for the interval density one obtains

$$\begin{aligned} p(\tau_i)\, d\tau_i &= (\sqrt{\pi}\sigma(\tau_i - r))^{-1} \\ &\times \exp\left[-\left(\mu - \tau^{-1} \ln \left(\frac{1 - r/\tau_i}{r/\tau_i} \right)\right)^2 \Big/ (\sigma^2/\tau) \right] d\tau_i, \end{aligned} \tag{64}$$

and for the variable u_i,

$$\begin{aligned} p(u_i)\,du_i &= (\sqrt{\pi}\sigma u_i)^{-1} \\ \text{(65)}\qquad &\times \exp\left[-\left(\mu + \frac{1}{\tau}\ln u_i\right)^2 \bigg/ (\sigma^2/\tau)\right] du_i; \end{aligned}$$

i.e., the variable

$$u_i = x_i^{-1} - 1 = \exp\left[-\tau^{-1}(\alpha v_i - c)\right]$$

has an equilibrium density that is *lognormal* (Kendall and Stuart [**13a**]), or alternatively, the function $v_i = \ln(1 - x_i/x_i)$ is normally distributed. Here then is a means for distinguishing between the predictions of the Gibbs-ensemble and the Langevin theories, in terms of the fit of the β-density of the second kind against the lognormal density for the variable u_i.

By similar arguments the linearized equations of the Langevin theory (Equation (48)) give rise to the equilibrium densities

$$\begin{aligned} \text{(66)}\qquad p(\tau_i)\,d\tau_i &= (\sqrt{\pi}\sigma)^{-1}(4r/\tau_i^2) \\ &\times \exp\left[-(\mu - \tau^{-1}(2 - 4r/\tau_i)^2/(\sigma^2/\tau))\right] d\tau_i, \end{aligned}$$

$$\begin{aligned} p(u_i)\,du_i &= (\sqrt{\pi}\sigma)^{-1}4(1 + u_i)^{-2} \\ \text{(67)}\qquad &\times \exp\left[-\left(\mu - \frac{2}{\tau}\left(\frac{1 - u_i}{1 + u_i}\right)\right)^2 \bigg/ (\sigma^2/\tau)\right] du_i \end{aligned}$$

$$\begin{aligned} p(v_i)\,dv_i &= p(y_i)\,dy_i \\ \text{(68)}\qquad &= (\sqrt{\pi}\sigma)^{-1} \exp\left[-\left(\mu - \frac{v_i}{\tau}\right)^2 \bigg/ (\sigma^2/\tau)\right] dv_i. \end{aligned}$$

As expected the variable v_i (which of course is the "canonical" variable of the Gibbs-ensemble theory) is again normally distributed. I ought to remark that the variable v_i is of course the membrane potential, suitably smoothed. Of course this result should not be taken too seriously, since the presence of a threshold for the emission of an impulse and the subsequent after-effects will evidently give rise to a nonnormal density for the actual membrane potential.

An alternative is to use the relationship given in Equation (2):

$$\tau \ln \frac{x_i}{1 - x_i} = c - \alpha \langle i \rangle \tag{2}$$

as obtaining in equilibrium. Then

$$v_i = (c/\tau) - (\alpha/\tau)\langle i \rangle \tag{69}$$

and both v_i and $\langle i \rangle$ are normally distributed. It is easy to show that any linear combination

$$I = \sum_k \alpha_k \langle i_k \rangle$$

has a density which is multivariate normal. This again has important consequences for a model relating EEG to single-unit activities. Similar remarks apply to the use of Equation (2) instead of Equation (59), with the Gibbs-ensemble.

4e. **Is there a "spectrum" of amplitudes of fluctuation of neural activities?** I now wish to conclude this paper on a somewhat speculative note. In discussing the connection between the Gibbs-ensemble and Langevin theories, I noted that a common equilibrium density is obtained when and only when one has in the Gibbs theory a specially coupled system of undamped harmonic oscillators, and in the Langevin case a collection of damped elements (or harmonic oscillators, for that matter). I conjectured that in the general case of a net of coupled nonlinear oscillators (with damping), the equilibrium density was of the form $\exp[-\beta V]$ where V is a Liapunov function of the system. What form is such a density likely to have for the nonlinearity that I have used in this paper? The answer I wish to propose is that the density will be multimodal, with as many modes as there are distinct *linear* regions in the nonlinearity. Consider now Equation (6), rewritten in the form:

$$(70). \quad x_i(t) = \varphi\left[\left(\tau \frac{d}{dt} + 1\right)^{-1}\left(\varepsilon_i + \frac{1}{\beta_i} \sum_j \alpha_{ij} x_j(t)\right)\right]$$

where $\varphi(y)$ is the inverse of $\ln(x/1 - x)$, i.e., the function

$$(71) \qquad \varphi(y) = \exp y(1 + \exp y)^{-1}.$$

It is clear that linear functions may be used to approximate this function whenever $|\varphi''(y)|$ is sufficiently small. Evidently on this criterion there are *three* regions where such an approximation will hold, corresponding to very low and very high cell sensitivities near one and zero, and an intermediate region of cell sensitivities centered about $x = 0.5$. Consider now the effects of a continued random bombardment of cells in a net, either by way of the many *interneurons* that play on them, or from miscellaneous extrinsic sources. Suppose this random input to be represented as a Gaussian noise input, i.e., δ-correlated white noise. It follows from § 3 that under some circumstances as cells are driven through their full range of activities, there will be a balance of activity (random input vs. damping) in just the three linear regions cited above, corresponding for some small δ to probability δ of response, probability $1 - \delta$ of response, and an intermediate level centered about the response probability of one half. Since $\varphi'(y)$ is greatest in this intermediate region and smallest in the two other regions, it follows that the equilibrium density should be *trimodal* with a wide central peak and two much narrower side peaks. The implication of this is that there should be three distinct phases in cell activities corresponding to different amplitudes of fluctuation, moreover by symmetry the amplitudes of fluctuation in the two extreme ranges should be equal. Thus if one were to record from cells of this type and make histograms of all the amplitudes of fluctuation recorded, the histograms should be *bimodal*.

I have attempted to test this prediction on some data collected by my colleague Professor A. D. J. Robertson on the responses of cells in the visual cortex to flashes of light. Figure 5 shows an example of the θ-histograms obtained from the data. (The ensemble is constructed by successively stimulating an animal with flashes of light at one second intervals. The amplitude of fluctuation is computed from spike counts in nine-millisecond-wide "bins" successively following the stimulus.) It will be

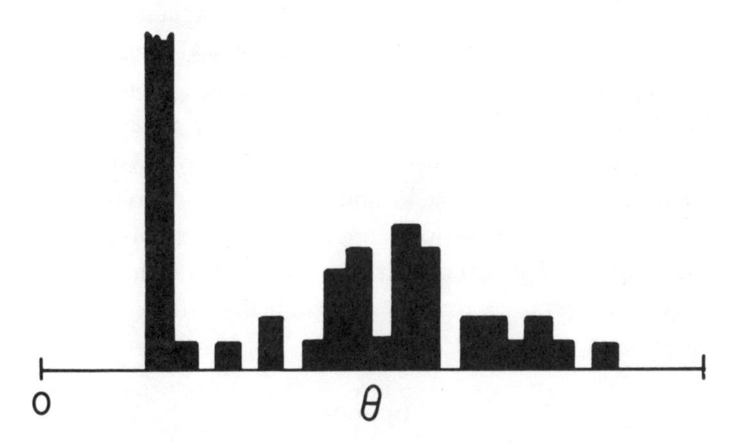

Figure 5
Histogram of the "amplitude of fluctuation" of neural activity (see Text).

seen that the histogram does have a structure that is consistent with my conjecture. In my opinion these findings are closely related to those of Smith and Smith [**25a**] who found that graphs of the logarithm of the distribution of interspike intervals showed two intersecting lines. Since a line on such a plot corresponds to a simple exponential point process (Cox [**5**]) with an interval density of $\lambda \exp(-\lambda\tau_i)$, it follows that an appropriate mathematical model of their data is a compound exponential process corresponding to the random firing of a cell in bursts of spike activity. In the model I have introduced in this paper bursts of spike activity correspond to oscillations and fluctuations in cell sensitivities and the conjecture concerning separate "states" in the activity is consistent with the data of Smith and Smith, at least for the two lower states. The reader is referred to the relevant experimental literature for the details.

The final point I wish to make about this is that here is the beginning of what might turn out to be a model for the distinct states of activity found in cortical cells which involves only such simple ideas as thresholds, circuits, and random excitations. As I have previously noted I feel that this kind of approach is a natural step beyond the McCulloch-Pitts approach, and it is therefore extremely interesting to observe what might be the signs of distinct states of activity in neural firing patterns-states that might serve as successors to the "all-or-none" activity of classical neurophysiology.

Appendix

(a) **The membrane potential.** The basic nature of the neuron has been determined by means of an extensive sequence of experiments in which extracellular and intracellular recordings were made of the electrical properties of the neuronal membrane (Eccles [7]). The neuron may be regarded as a globule bounded by the surface membrane, and containing a salt solution very different in composition from the external medium. All changes in the membrane potential are attributed to ionic movements across the surface membrane. Figure 6 illustrates the manner of operation of an inhibitory and an excitatory synapse. (The locus at which impulses pass from the axon of one cell to the somato-dendritic membrane of another.) The ionic current flow across the membrane is inward at an excitatory synapse, outward at an inhibitory synapse. The resting

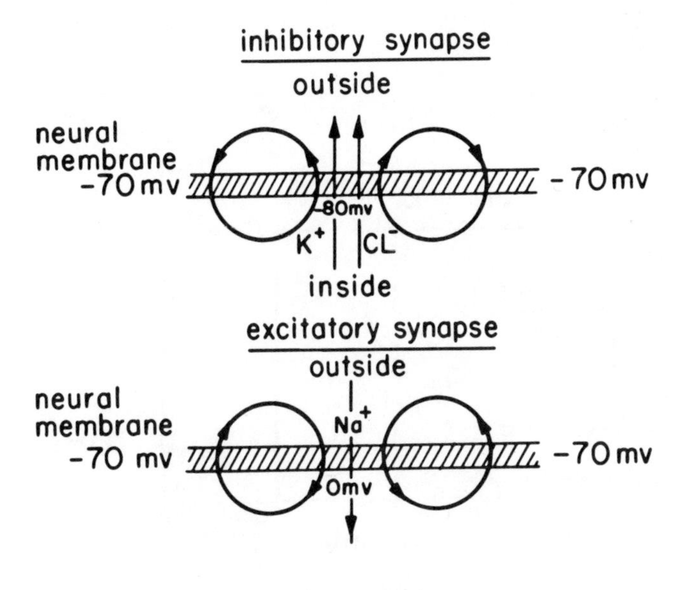

FIGURE 6
Current flows at an inhibitory and at an excitatory synapse.

membrane potential is about -70 mv, the equilibrium potential for excitation is 0 mv, and that for inhibition is -80 mv. An equivalent electrical circuit which gives the impedances and sources is shown in Figure 7. The membrane resistance R_m represents the resistance of that part of the somato dendritic membrane which is not covered by synapses. It is therefore independent of subsynaptic currents and is normally about 1 MΩ. The time-constant of the membrane, τ_m, is about 3 msec. It is obtained by measuring the time-course of the membrane response to unit current steps. The membrane capacitance is then $C_m = g_m/R_m$, about 3 mμF. The variable resistances in the equivalent circuit represent the manner of operation of the excitatory and inhibitory subsynaptic areas of the membrane. In the absence of presynaptic excitation these resistances are effectively infinite; i.e., there is an open circuit between

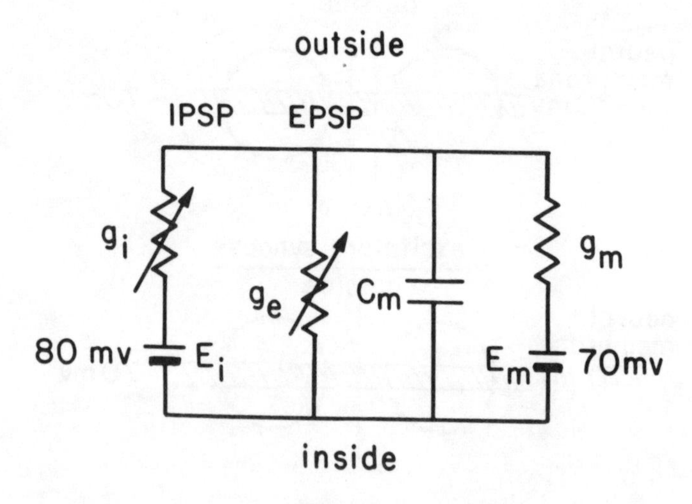

FIGURE 7
"Lumped equivalent-circuit" for the somato-dendritic neural membrane.

presynaptic axons and the secondary cell membrane. During pre-synaptic excitation the inhibitory synaptic resistance may drop to about 0.7 MΩ, the excitatory resistance to about 0.5 MΩ. The responses to presynaptic excitation are called respectively, "inhibitory post synaptic potentials" or ipsps, and "excitatory post-synaptic potentials" or epsps. Generally the maximum rate or rise of either an epsp or an ipsp at a single synapse is about 1 v/sec.

These phenomena are conveniently expressed by the differential equation

$$\text{(1A)} \qquad \left[\tau\frac{d}{dt} + 1\right]v(t) = \frac{\tau}{C_m}\,[I_m + (-E_m)/R_E + (E_I - E_m)/R_I].$$

$v(t)$ is the membrane potential measured relative to the resting potential $-E$ and is therefore $v_{\text{inside}} - v_{\text{outside}} - E$. τ is the time-constant of the membrane. It is given by the expression

$$\text{(2A)} \qquad \frac{1}{\tau} = \frac{1}{C_m}\left[\frac{1}{R_m} + \frac{1}{R_E} + \frac{1}{R_I}\right].$$

When the synapses are open-circuited the time constant is then $\tau_m = R_m C_m$. The current I_m is simply an applied current flowing across the membrane. From these equations it will be seen that the neural membrane is a linear time-varying system, so far as its passive sub-threshold behavior is considered. The temporal variation is seen to be caused by variation in the synaptic resistances R_E and R_I each of which may be thought of as the resultant of the parallel combination of numerous individual synaptic resistances, each in series with a switch which may be closed by presynaptic excitation. Suppose the system is initially at the resting potential so that $v(0) = 0$, and $R_E = R_I = \infty$. Suppose that the excitatory synaptic resistance r_i is "switched-on" at $t = 0$. The solution of Equation (1A) for v at time t' is then

$$v(t') = [\tau/C_m][-E_m/r_i](1 - \exp(-t'/\tau)),$$

$$\frac{1}{\tau} = \frac{1}{C_m}\left[\frac{1}{R_m} + \frac{1}{r_i}\right].$$

If r_i now switches off at t', the membrane potential at time t is given by

$$v(t) = v(t') \exp(-(t - t')/\tau_m).$$

Thus the membrane potential builds up exponentially toward the equilibrium value $[\tau/C_m] \cdot [-E_m/r_i]$ as long as the resistance r_i is switched on, and decays exponentially towards the resting potential $v = 0$ whenever r_i is switched off. Suppose now that r_i is switched on for an interval δt, then

$$v(\delta t) = [\tau/C_m][-E_m/r_i]\left(\frac{\delta t}{\tau} + O\left(\frac{\delta t}{\tau}\right)^2\right).$$

r_i is the resistance at one excitatory synapse. It is larger than R_E by several orders of magnitude. Therefore $\tau - \tau_m$. The synaptic operating time δt is about 1 msec so that $O(\delta t/\tau)^2$ is about 10% of $\delta t/\tau$. It is therefore appropriate to take

$$\frac{\delta t}{\tau} + O\left(\frac{\delta t}{\tau}\right)^2 = \alpha\frac{\delta t}{\tau},$$

where $a = 1.1$ is a "fudge-factor". Then

$$v(\delta t) = [\alpha\, \delta t/C_m] \cdot [-E_m/r_i].$$

Let $v(\delta t) = \delta v_{ij}$, $\alpha\, \delta t = \sigma$, $1/r_i = \delta g_{ij}$, and $-E_m = E_j$, then the expression given in § 1 is obtained for δv_{ji}, the deviation in the membrane potential of the ith secondary cell caused by the arrival of an impulse from the jth primary cell.

Let $Q_i(t)$ be the excess charge in the ith cell at time t. Since $Q_i(t) = C_m v(t)$ it is evident that $\delta Q_{ij}(t)$, the variation in Q_i caused by the arrival of an impulse, is simply $C_m\, \delta v_{ij}(t)$. It is assumed that δQ_{ij} is delivered impulsively at time t; i.e., that the finite amount of charge δQ_{ij} is placed across the cell membrane at the instant t. This is equivalent to taking the action potential to be the Heaviside-Dirac impulse function. The membrane response to such an impulse is evidently

$$h_{ij}(t) = u_{-1}(t)[\delta v_{ij}] \exp [-t/\tau_m].$$

It is now assumed that the membrane response to different synaptic imputs is *linear*. Let $F_j(t)$ be the distribution function of the times of arrival of impulses from the jth cell. Then the potential caused by impulses from N primary cells is given by the Stieltjes integral

$$v_i(t) = \sum_{j=1}^{N} \int_0^t h_{ij}(t - \tau)\, dF_j(\tau).$$

This may be rewritten as the convolution

$$v_i(t) = \sum_{j=1}^{N} \int_0^t h_{ij}(t - \tau) f_j(\tau)\, d\tau \tag{3A}$$

where $f_j(t) = dF_j/dt$ is the mean frequency of arrival of impulses from the jth cell. On the assumption that $f_j(t)$ is slowly varying relative to the time-constant τ_m,

$$v_i(t) = \sum \tau_m\, \delta v_{ij} f_j(t)[1 - \exp(-t/\tau_m)],$$

or alternatively

$$v_i(t) = \sigma \sum_j \left(\frac{\delta g_{ij}}{g_m}\right) E_j f_j(t)[1 - \exp(-t/\tau_m)],$$

where $g_m = 1/R_m$ is the resting membrane conductance. Strictly speaking the membrane conductance at time t is the linear combination

$$\begin{aligned} g &= g_m + \sigma \sum_j \delta g_{ij} f_j(t) \\ &= g_m \left[1 + \sigma \sum_j \left(\frac{\delta g_{ij}}{g_m}\right) f_j(t)\right] \end{aligned}$$

and the membrane time-constant is

$$\tau = \tau_m \left[1 + \sigma \sum_j \left(\frac{\delta g_{ij}}{g_m}\right) f_j(t)\right]^{-1}.$$

On the assumption that there are approximately 10,000 synapses per cell each with a conductance δg_{ij} of 0.2 mμ mho, and that $f_j(t)$ is approximately 100 impulses per sec., $\sigma \sum_j (\delta g_{ij}/g_m) f_j(t) \sim 0.2\tau_m$. Once again fudgefactors b and c are used to express these deviations. Thus

$$\text{(4A)} \quad v_i(t) = \sigma \sum_j \left(\frac{\delta g_{ij}}{g_m}\right) E_j f_j(t)(1 - \exp(-t/C\tau_m))$$

where b is incorporated in σ, so that $b = c = 1.2$ and $\sigma = 0.92\ \delta t$. These approximations effectively reduce the linear time-varying system which has been considered, to a linear time-invariant system, the response of which is given by Equation (4A). Recent calculations by Rall [**23**] indicate that the "lumped" equivalent circuit underestimates the membrane time-constant by perhaps a factor of 2. In cells such as pyramidal neurons which have relatively thick apical dendrites there may be appreciable delays and attenuations at the soma, of peripheral changes in the membrane potential. In part this is due to the varying geometry of the neuronal membrane, and to the fact that the membrane impedances and sources are "distributed" over a membrane approximately 0.5 to 1 mm long. The equation representing such a situation is not Equation (1A), but the telegraphist's equation for a lossy noninductive cable of diameter d,

$$\lambda^2 \frac{\partial^2 v}{\partial x^2} = \tau \frac{\partial v}{\partial t} + v,$$

where x is distance from the soma and where λ is the space-constant of the membrane, proportional to $\sqrt{d}$. Rall ([**23**] *et seq.*) has used this equation to represent branching dendritic trees, in which case the problem involves multiple region boundary values. The results clearly indicate that the system is not only nonstationary in the time-constants λ and τ, but also nonlinear in its responses to multi-region conductance changes caused by synaptic activity. Superposition is a rare property of such a system, and both the temporal sequence and the locus of synaptic activity are important determinants of the membrane response. The first-order linear approximation neglects both the geometry and the temporal variation of the membrane constants.

(b) **Supra-threshold behavior.** When the membrane potential reaches a certain threshold value θ, changes occur in the membrane conductance g_m which result in the generation of the action potential. The action potential is a rapid change of the membrane potential from the threshold value $\theta \sim 15$ mV to a value of around $v = 100$ mV, followed by a gradual return to a sub-threshold value. As is well known the Hodgkin-Huxley equations adequately represent the sequence of changes in g_m and the resulting action potential (at least for axons), (Hodgkin and Huxley [**12**]). However these equations are equivalent to a fourth-order nonlinear differential equation with time-varying coefficients which has not been solved in closed form. It is therefore necessary to approximate these nonlinear and active membrane characteristics by somewhat simpler and mathematically tractable forms.

In this paper, instead of attempting to model the events giving rise to a single action-potential, this potential is represented simply as a delta-function, and only the relationship between the mean *rate* at which a cell emits such impulses and the voltages and currents driving the cell is utilized. The exact functional relationship used is given in Equation (2)

$$\tau \ln \frac{x_i}{1 - x_i} = c - \alpha \langle i \rangle \tag{2}$$

or from Equation (59)

$$\tau \ln \frac{x_i}{1 - x_i} = c - \alpha \langle v \rangle . \tag{59}$$

Equation (2) is a frequency-current relationship whereas Equation (59) relates cell discharge frequency to membrane voltage. Of course Equation (59) follows from Equation (2) by way of the special dynamics assumed in this paper. It is likely that more realistic models of neural dynamics will give rise to mixed relationships with discharge frequencies related to both currents and voltages across the membrane. I want to note here that published data of Creutzfeldt *et al.* [**6**] using intracellular microelectrodes in cortical cells fits quite well Equation (2), and that unpublished data obtained by my colleague Robertson using extracellular microelectrodes on cortical cells fits quite well *both* relationships, i.e., some cells are best approximated by Equation (2), some by Equation (59).

Acknowledgements

The work reported in this paper has been carried out over the last few years while I was a visitor at the Imperial College of Science and Technology, London, at the National Physical Laboratory, Teddington, and at the Laboratorio di Cibernetica, C.N.R., Naples. I am grateful for the hospitality extended me by these institutions. I have had many useful discussions with many persons, too many to cite individually, but I should like to pay special tribute to my colleague Professor Morrel Cohen at the University of Chicago, and to Dr. Peter Johannesma of the Catholic University of Nijmegen for their very helpful comments and criticism. I should also like to acknowledge the most efficient services of Miss Anita Larsson and Mrs. Dorothy Cooney in the production of this manuscript.

Finally this work has been partly supported by the Physics Branch of the Office of Naval Research, U.S.N. I am truly grateful for this support.

REFERENCES

1. P. Andersen and S. Andersson, *The physiological basis of the alpha rhythm*, 1968.

2. G. D. Birkhoff, *Proof of a recurrence theorem for strongly transitive systems; proof of the ergodic theorem*, Proc. Nat. Acad. Sci. U.S.A. **17** (1931), 650–660.

3. R. L. Butchart, *An explicit solution to the Fokker-Planck equation for an ordinary differential equation*, Internat. J. Control (1) **1** (1965), 201–208. MR **33** #1173.

4. R. Cajal, *Histologie du system nerveux*, C.N.I.C. Madrid, 1952.

5. D. R. Cox, *Renewal theory*, Methuen, London and Wiley, New York, 1962. MR **27** #3030.

6. Creutzfeldt, et al, *Brain and conscious experience*, Springer-Verlag, New York, 1964.

7. J. C. Eccles, *The physiology of nerve cells*, Johns Hopkins Press, Baltimore Md., 1957.

8. ——— *The physiology of synapses*, Springer-Verlag, New York, 1964.

9. J. C. Eccles, et al, *The cerebellum as a neuronal machine*, Springer-Verlag, New York 1967.

10. W. Feller, *An introduction to probability theory and its applications.* Vol. 2, Wiley, New York, 1966. MR **35** #1048.

11. J. W. Gibbs, *Elementary principles in statistical mechanics*, Yale Univ. Press, New Haven, Conn., 1902.

12. A. L. Hodgkin and A. D. Huxley, *A quantitative description of membrane current and its application to conduction and excitation in nerve*, J. Physiology **117** (1952), 500–544.

13. P. I. M. Johannesma, *Neural networks*, Springer-Verlag, New York, 1969.

13a. Maurice G. Kendall and Alan Stuart, *The advanced theory of statistics*, Vol 1: *Distribution theory*, Hafner, New York, 1958. MR **23** #A2247.

14. E. H. Kerner, *A statistical mechanics of interacting biological species*, Bull. Math. Biophys. **19** (1957), 121–146. MR **19,** 374.

15. ———, *Further considerations on the statistical mechanics of biological associations*, Bull. Math. Biophys. **21** (1959), 217–255. MR **21** #3278.

16. ———, *On the Volterra-Lotka principle*, Bull. Math. Biophys. **23** (1961), 141–157. MR **23** #B3047.

17. M. Kimura, *Diffusion models in population genetics*, J. Appl. Probability **1** (1964), 177–232. MR **30** #2946.

18. R. Kubo, *Statistical-mechanical theory of irreversible processes. I: General theory and simple applications to magnetic and conduction problems*, J. Phys. Soc. Japan **12** (1957), 570–586. MR **20** #4940 a.

19. E. G. Leigh, Jr., *The ecological role of Volterra's equations*, Lectures on Math. Life Sciences, vol. 1, Amer. Math. Soc., Providence, R.I., 1968, pp. 1–14.

20. A. J. Lottka, *Elements of theoretical biology*, Dover, New York, 1924; reprint 1956.

21. W. S. McCulloch and W. Pitts, *A logical calculus of the ideas immanent in nervous activity*, Bull. Math. Biophys. **5** (1943), 115-133. MR **6,** 12.

22. R. Pearl, *The growth of populations*, Quarterly Rev. Biology **2** (1927), 532–548.

23. W. Rall, *Distinguishing theoretical synaptic potentials computed for different soma-dendritic distributions of synaptic input*, J. Neurophysiology **30** (1967), 1138–1168.

24. S. A. Rice and P. R. Gray, *The statistical mechanics of simple liquids. An introduction to the theory of equilibrium and non-equilibrium phenomena*, Interscience, New York, 1965. MR **35** #5203.

25. A. D. J. Robertson, Personal communication, 1969.

25a. D. R. Smith and G. K. Smith, *A statistical analysis of the continual activity of single cortical neurones in the cat unanaesthetized isolated forebrain*, Biophysical J. **5** (1965), 47–74.

26. R. B. Stein, *The frequency of nerve action potentials generated by applied currents*, Proc. Roy. Soc. Ser. B. **167** (1967), 64–86.

27. V. Volterra, *Leçons sur la théorie mathématique de la lutte pour la vie*, Gauthier-Villars, Paris, 1931.

28. N. Wax, *Selected papers on noise and stochastic processes*, Dover, New York, 1954.

29. N. Wiener, *Time, communication and the nervous system*, Ann. New York Acad. Sci. **50** (1948), 197–220. MR **10**, 133.

30. R. Zwanzig, *Memory effects in irreversible thermodynamics*, Phys. Rev. (2) **124** (1961), 983–992.

31. I. Prigogine, *Non-equilibrium statistical mechanics*, Monographs in Statistical Physics and Thermodynamics, vol. 1, Interscience, New York, 1962. MR **32** #5286.

GRAPHICAL ANALYSIS OF ECOLOGICAL SYSTEMS

By

ROBERT Mac ARTHUR[1]

Princeton University

[1] The work presented here has been supported by grants from the National Science Foundation.

1. Introduction

My aim is very simply stated: I want to learn as much as possible by graphical means about the solutions of the "ecological equations"

$$
\begin{aligned}
\frac{dX_1}{dt} &= X_1 f_1(R_1, R_2, \ldots) \doteq X_1[a_{11}R_1 + a_{12}R_2 + \cdots - T_1] \\
\frac{dX_2}{dt} &= X_2 f_2(R_1, R_2, \ldots) \doteq X_2[a_{21}R_1 + a_{22}R_2 + \cdots - T_2] \\
&\;\;\vdots \\
\frac{dR_1}{dt} &= R_1 g_1(X_1, X_2, \ldots, R_1, R_2, \ldots) \doteq c_1 R_1[K_1 - b_{11}X_1 \\
&\qquad\qquad - b_{12}X_2 - \cdots - R_1] \\
\frac{dR_2}{dt} &= R_2 g_2(X_1, X_2, \ldots, R_1, R_2, \ldots) \doteq c_2 R_2[K_2 - b_{21}X_1 \\
&\qquad\qquad - b_{22}X_2 - \cdots - R_2] \\
&\;\;\vdots
\end{aligned} \tag{1}
$$

Here the X's are the populations of consumer species and the R's of their resources; T_i is the threshold quantity of resource which will allow X_i to rear its young; in the absence of consumers, resource R_i will asymptotically reach level K_i. The right-hand column appeals to ecologists as "about right" as an initial guide to the form the functions f_i and g_i should have, but no two systems will have exactly the same functional form, let alone the same values of the constants, and we want to make sure

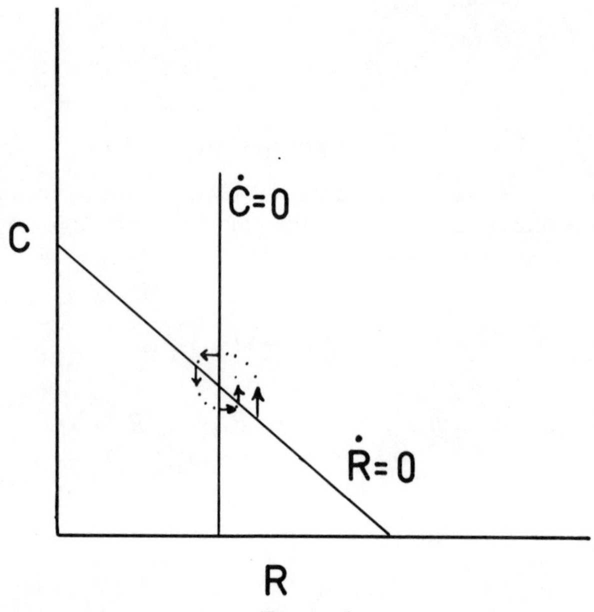

FIGURE 1

Isocline analysis of consumer-resource systems. $\dot{C} = 0$ is the left boundary of the region in which consumers C increase and $\dot{R} = 0$ is the upper boundary of the region in which the resources R can increase. The arrows, and, roughly, the dots, indicate the history of the populations as predicted from the graph. This is locally stable.

that any conclusions drawn about the solution are not artifacts of the exact form postulated for these functions. This is the great virtue of graphical analysis, but the fact that we can only draw in two dimensions also imposes some limitations. I shall begin with the graphical analysis of the precise equations and then let the functions bend in reasonable ways and see how the predictions are affected.

2. The dynamics of a consumer-resource system

I shall first peel off one consumer species C and one resource R and discuss their dynamics, following Rosenzweig and Mac Arthur [**3**]. Beginning with the precise

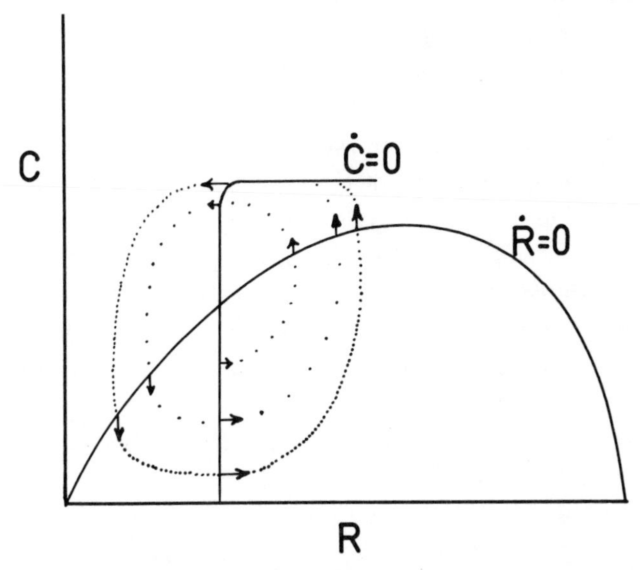

FIGURE 2
As Figure 1 except that the isoclines are bent to include a minimum population for the resources and a ceiling population for the consumers. The positive slope, at the intersection, of the R isocline causes a local instability as shown by the outward spirals. The consumer ceiling stops these increasing oscillations and produces a limit cycle of an undetermined shape but conceivably as shown by the closely spaced dots.

eight-hand versions of Equations (1), we plot the regions in which $dx/dt > 0$ and $dR/dt > 0$. These are shown in Figure 1 and the arrows indicate the direction of motion of the system which, it can be shown, spirals inward. Now, as my student Maly has shown experimentally, the resource isocline $dR/dt = 0$ actually may bend down at the left as in Figure 2, and now the consumer isocline can intersect it in its rising part which will cause outward spiralling arrows; i.e., it is locally unstable. Maly has shown that at least one laboratory system is of this type, which helps explain the notorious difficulty in getting consumer and resource to coexist in the laboratory.

Furthermore, a small time lag τ, in the responses of either consumer or resource, will accentuate the instability and may even cause trouble in systems like Figure 1. (The arrows point in the direction they would have pointed a time τ ago and radiate outward.)

Two simple features of the system can give it a global stability or cause limit cycles, however. First, if there is a ceiling on consumer abundance, set perhaps by its own territoriality, then the C isocline bends horizontal as in Figure 2. Now the arrows spiral outward until they hit the horizontal part of the C isocline and from the time they leave this isocline they are in a limit cycle, for the next cycle will also leave the isocline at the same point and hence follow the same path.

Adding a hiding place for the resource species can, under certain circumstances, bend the resource isocline vertical at the left and the same reasoning shows that outward spirals will reach a limit cycle.

It is possible to draw three dimensional graphs of carnivore, herbivore and plant, two consumers and one resource or one consumer affecting the competition of two resources, but that is about the limit of the complexity that can be dealt with by the conventional phase plane analysis.

3. The statics of a two resource-many consumer system

If we abandon the goal of studying the changes in population with time, we can proceed to much more complicated systems and still gain an understanding of the equilibria. I plot the X isoclines $dX_i/dt = 0$ in R_1 and R_2 coordinates. This is useful whenever the functions f_i have no X's in their arguments and, in this case, implies that the

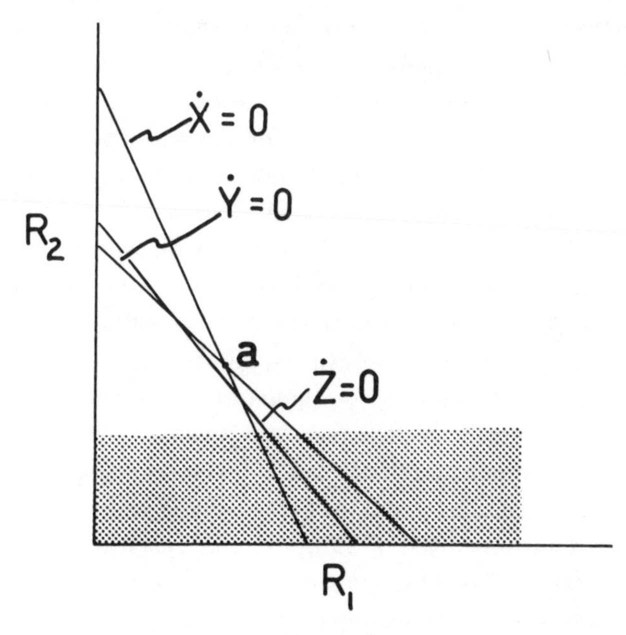

FIGURE 3

The inner boundaries of the resource levels for which $\dot{X} > 0$, $\dot{Y} > 0$ and $\dot{Z} > 0$ are shown as lines, not necessarily straight. If X and Y come to equilibrium, it must be where the resources are at the levels determined by point a. In this figure $\dot{Z} > 0$ at resource level a, and so Z can invade. The stippled area, with upper right corner at coordinates K_1, K_2 limits the attainable resources. In this environment the X isocline is innermost and X alone will be found. For regions with different K_1 and K_2 different species combinations may be found.

consumers are limited by their resources but not by any aggression or territoriality. With this assumption I now plot three X_i isoclines in the R_1, R_2 plane as shown in Figure 3. Notice that $(dX_i/dt) > 0$ outside of the isoclines rather than inside. Now there may not be an equilibrium of the various X's, but if there is, it must lie where all isoclines meet. A glance at the graph shows that only pairs of the isoclines intersect (it would be triples in a three resource system, and so on) so that *at most two of these consumers can persist at equilibrium*. But which two? If X and Y come to equilibrium, the resources will be at level a, and clearly at that level of the resources Z, can still increase. That is, Z can invade. On the other hand, the systems X and Z or Y and Z at equilibrium are neither invasible by the third species. In fact, they may be alternate stable equilibria. But all this discussion has been predicated on the existence of the equilibria. How can we tell whether, in fact, the equilibrium is attainable?

A necessary condition for the equilibrium can be discussed graphically as in Figure 3 by examining the stippling. R_1 cannot exceed K_1 and R_2 cannot exceed K_2, so these are a generous estimate of attainable R values and we are sure no unstippled R values can be reached. Hence our system will be confined within this stippled rectangle and, in this particular example, no two-species equilibrium is in fact possible. Rather, the innermost isocline, which is owned by X, lies wholly inside of the resource levels at which Y and Z can maintain their populations. X can always invade and replace the others. Of course, in a different environment, K_1 and K_2 will generally be different and if the new stippled area includes the intersection of two isoclines, equilibrium may be possible. In fact, we may draw biogeographic conclusions and predict what sequence of species will be encountered as we progress across a landscape which changes from $K_1/(K_1 + K_2) = 0$ to $K_1/(K_1 + K_2) = 1$. Stippling in such areas, we have various alternatives. If $K_1 + K_2$ is a constant, small, value, X might first occur, followed by an empty space, followed by Y. If $K_1 + K_2$ is larger, then the sequence would be Y, $Z + Y$, $Z + X$, X. If there were four

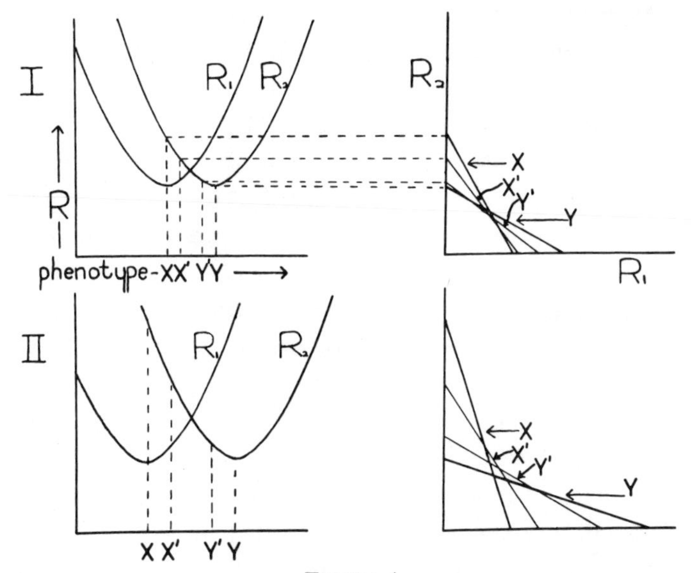

FIGURE 4
The right-hand graphs are like that of Figure 3, and show, in the upper case, X' and Y' invading and replacing X and Y. In the lower case X' and Y' are invaded and replaced by X and Y. The left-hand sides show the R_1 and R_2 intercepts, as continuous functions of the measurements (phenotype) which can be made on species. Thus X and Y in both cases are the phenotypes which can reduce resources R_1 and R_2 respectively to lowest levels. But in graph II, the resources are more different and X', Y' diverge to X and Y; in graph I, the resources are more similar and X and Y converge to and beyond X', Y'. Thus, by using the right-hand graphs to interpret the left-hand ones, we can show both convergence and divergence.

instead of three species with X_4 even better on R_2 we would get X_4, $X_4 + X_3$, $X_3 + X_2$, $X_2 + X_1$, X_1. This illustrates the general features of the system: no more species than resources will exist in any place and the ones which coexist are segments from the ranked arrangement of the species. That is, we do not get X_4 and X_1 together. The species whose ranges abut are not those that are the most severe competitors.

There is a way of using these graphs to predict evolutionary convergence and divergence. The intercepts in these graphs are assumed to be continuous functions of the morphology of the organisms. Morphology is multidimensional, but I will concentrate on some single dimension. In the left-hand graphs of Figure 4, the intercepts on the R_1 and R_2 axes are plotted as U shaped curves against morphology. Four separate morphologies, X, X', Y', Y, are singled out for attention and their isoclines are plotted, by joining the intercepts with straight lines, in the right-hand figures. In the upper graph the X' and Y' isoclines intersect inside the point where X and Y intersect; i.e., X' and Y' replace X and Y. Thus there would be favored an evolutionary convergence of X and Y. In the lower graphs, which are different only in that the U shaped curves are farther apart, meaning that the resources are more different, the X' and Y' isoclines intersect farther out than the X and Y proving that divergence of X' and Y' is favored. This kind of analysis can be extended to the case of a continuous spectrum of resources (Mac Arthur and Levins [**1**]).

4. Quasi equilibrium

I can proceed one stage farther with the analysis by yet another kind of graph. Again I shall plot the X_i isoclines but this time in a graph whose coordinates are K_1 and K_2. (These are determined experimentally by excluding consumers from some resource and seeing to what level they rise in the absence of consumers.) Now, to do this, I must make additional restrictions, since K's do not appear in the dX/dt equations. I make an assumption reminiscent of thermodynamics books in which the authors say they deal with systems which as they change differ only by infinitesimals from equilibrium states. Here I restrict consideration to systems in which the R values and X values always are very close to the equilibrium values. To make the mathematics neater, I write Equations (1) (using the precise right-hand sides) in matrix form, $\dot{X} = X[AR - T]$, where R and T are column vectors, X is a row vector and A is a matrix of a_{ij}. Similarly, $\dot{R} = R[K - BX - R]$, where B is the matrix of b_{ij}, K and X are now column vectors and R is now a row vector except inside the bracket. We happily identify vectors with their duals so that rows and columns are descriptions of the same thing. Now near equilibrium $\dot{R} = 0$, so $K - BX = R$. Substitute this R into the upper equation,

$$\dot{X} = X[(AK - T) - ABX]. \tag{2}$$

This is the vector form of the Volterra competition equations and shows how they can be derived from the consumer-resource equations (always assuming the consumers are resource limited). Moreover, the derivation decomposes the coefficients of the competition equations into the much more natural coefficients A, B, K, and T of the former equations, and thereby gives us a recipe for calculating these constants. However, these difficulties are not the subject here; I can now plot $dX_i/dt = 0$ in the K_1, K_2 coordinates. In fact, $dX_i/dt = 0$ for all i if $ABX = AK - T$. Solving by Cramer's rule, X_i is determined as a ratio of determinants, the one in the numerator having the column vector $AK - T$ in its ith column. But determinants are linear, so this is K_1 times one number plus K_2 times another minus a third number. In other words, the numerator for any X_t is a linear function of the K's.

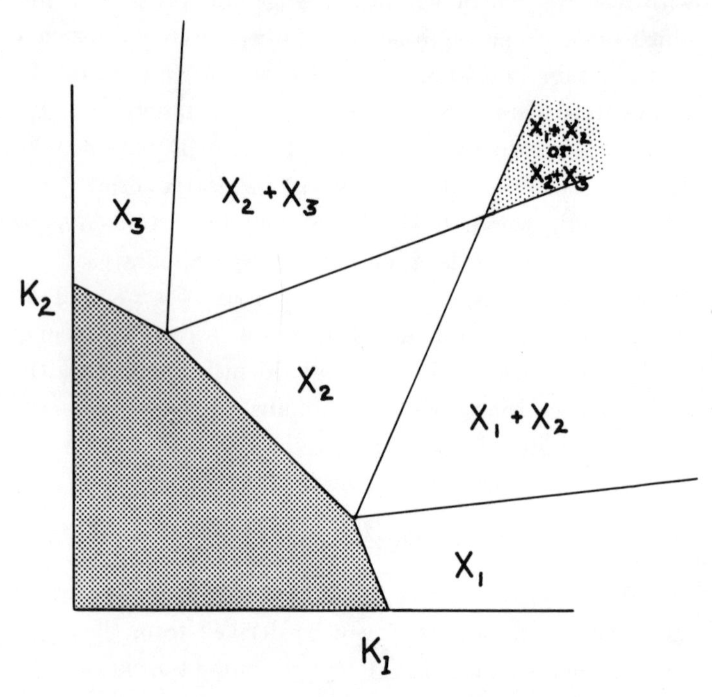

FIGURE 5
In the densely stippled region, no species can live. When the K values (asymptotic levels of resources in absence of consumers) are larger, at most two of the consumers, X can coexist, as indicated for a three-species system X_1, X_2, X_3. The lightly stippled region is the zone of alternate stable states where either X_1 and X_2, or X_2 and X_3, whichever gets there first, will be found. In the absence of competitors each species will expand to occupy all regions beyond its own line. For instance, X_1 will expand to occupy all regions to the right of the *extended* line that separates the dense stippling from the area marked X_1. More realistically the lines will bend somewhat.

In the case where B is independent of K, Figure 5 shows the nature of coexistence. The zone of expansion here, where X_1 can live only if X_2 is missing, and conversely, is one of the most testable aspects of the whole theory. Island species released from mainland competitors often expand their habitats greatly (Mac Arthur and Wilson [2]). From this graph, most of the conclusions drawn in § 3 can be made quantitative. The B matrix often depends upon the K's which causes the lines in Figure 5 to bend somewhat.

Since this was first written I have realized that if b_{jk} takes the probable form $b_{jk} = (K_j/r_j)a_{kj}$, then the equilibrium described by equation 2 is also the set of values which minimize a certain quadratic form. In particular, if $T_i = T$ and $\sum_j a_{kj} =$ constant so that $S_j = Tr_j/\sum_j a_{kj}K_j$ is independent of k, then

$$Q = \sum_j \frac{K_j}{r_j}(r_j - s_j - x_1 a_{1j} - x_2 a_{2j} - \cdots)^2$$

is minimized at equilibrium. This gives another graphical analysis of equilibrium, for the minimizing of Q can be viewed as a least-squares approximation of the r_j by sums of a_{kj} terms. (For more details, see Mac Arthur [4].)

5. Effects of fluctuating environment

So far, the equations have not involved time explicitly, but there are some situations when seasonality has interesting effects. I shall touch on them here. If $dX/dt = Xr(t)$, then $X(t) = X_0 \exp[\int_0^t r(t)\,dt]$. I ask what constant value $\bar{r}$ of $r(t)$ would yield the same value of $X(t)$ and find $\bar{r} = (1/t)\int_0^t r(t)\,dt$, so that $\bar{r}$ is the time average of $r(t)$. Now if one population has a larger $\bar{r}$ than another, it will eventually predominate. This is the mechanism both of competition and of natural selection.

Now suppose $r(t)$, which is the excess of per capita birth rate over death rate, is a function, say, of a fluctuating supply of resource G. Then $r(t) = r(G(t)) = r(\bar{G}) + (G - \bar{G})r'(\bar{G}) + (G - \bar{G})^2(r''(G^*)/2)$, where G^* lies between

$\bar{G}$ and G, according to Taylor's theorem. Therefore

$$\bar{r} = \frac{1}{t}\int_0^t r(t)\,dt = \int_0^\infty r(G)p(G)\,dG$$

where $p(G)$ is the proportion of time during which the resource has value G, and this last integral becomes roughly

$$\bar{r} = r(\bar{G})\int_0^\infty p(G)\,dG + r'(\bar{G})\int_0^\infty (G-\bar{G})p(G)\,dG + \frac{r''(G^*)}{2}\int_0^\infty (G-\bar{G})^2 p(G)\,dG$$

$$= r(\bar{G}) + \frac{r''(G^*)}{2} \text{ variance } G.$$

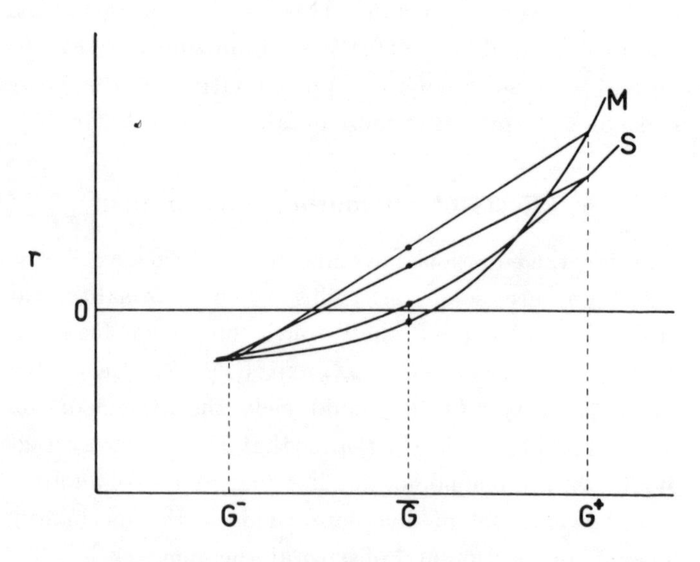

FIGURE 6
Two hypothetical species, M and S are considered with very different response curves of r (birth rate minus death rate) to food supply. M will outcompete S in a fluctuating food supply (varying from G^+ to G^- with mean $\bar{G}$), while S will outcompete M in a uniform density $\bar{G}$ of food.

Hence if the variance in G is large and the slope of $r(G)$ is uniformly increasing, the second term will make substantial increase in $\bar{r}$.

This whole matter is simpler in a graphical presentation as in Figure 6. Here species M can outcompete species S if G has a large variance, while S can win if G has a small variance. The intriguing question remains: Under what conditions is $r''(G^*)$ large and positive, as for species M in the figure, and when is it not? The answer requires a more elaborate analysis, but it seems to me that this is the mechanism behind both the large litter size and the large body size which are regularly found in seasonal climates.

REFERENCES

1. R. Mac Arthur and R. Levins, *The limiting similarity, convergence and divergence of coexisting species*, Amer. Naturalist **101** (1967), 377–385.

2. R. Mac Arthur and E. O. Wilson, *The theory of island biogeography*, Princeton Univ. Press, Princeton, N.J., 1967.

3. M. Rosenzweig and R. Mac Arthur, *Graphical representation and stability conditions of predator prey interactions*, Amer. Naturalist **97** (1963), 209–223.

4. R. Mac Arthur, *Species packing and what interspecies competition minimizes*, Proc. Nat. Acad. Sci. **64** (1970), 1369–1371.

EXTINCTION

By

RICHARD LEVINS

University of Chicago

It is a commonplace of evolutionary biology that the fate of most species which ever existed has been extinction. Since there are now roughly 10^6 species of plants and animals, and since this is less than 1% of the species which have existed, some 10^8 or 10^9 species extinctions have occurred in the 10^8–10^9 years since the start of the Cambrian. Thus over the whole of geological time species have been disappearing at the rate of roughly one per year.

Despite such a large number of cases, extinction has usually been regarded as a unique event. The extinctions of whole groups have been noted. Extinct intermediate ancestors have been postulated. Explanations have been offered for the extinction of particular groups, but no quantitative theory of extinction has been attempted.

Our basic proposition is that a species can be regarded as a population of populations and the extinction of a species is identical with the extinction of all its constituent populations. Therefore we approach the study of extinction of species by way of the dynamics of the extinction and founding of local populations.

Mac Arthur and Wilson [**1**] have used a model of balanced extinction and migration to build a theory of island biogeography and offer data to indicate that there is an appreciable turnover of species. Wilson and Simberloff [**3**] found for tiny islands in the Florida Keys that the local extinction rate approximates one species every few days. Harold Heatwole and I, studying the islands of the

Puerto Rico Bank, found at least a dozen extinctions of ant species on Cayo Ahogado in the course of five years, although the average fauna contained 2 or 3 species. The next step is the assertion that the distribution of many species even on the mainland is insular. Mountain tops, lakes, individual host plants, a fallen log, a patch of vegetation, a mammalian gut, or, less obviously, a region of optimal temperature or humidity are all islands for the appropriate organisms. Therefore the insular model is much more broadly applicable.

If populations are frequently being established and eliminated, there is the likelihood that the genetic composition of the species will change as a consequence. All that is required is that the probability of extinction of a whole population be a function of its gene frequency. This interpopulation selection has been invoked to account for many traits which do not seem to confer any advantage on their bearers, or which may even be disadvantageous (Wynn-Edwards [**4**]), and is at least part of the rationale for the use of optimization methods in evolutionary ecology. Williams [**2**] criticizes the use of group selection in evolutionary theory by invoking Occam's razor. He claims that all cases previously explained by group selection (including interpopulation selection) can be accounted for by the more familiar mendelian selection. However, the issue is not whether we need population selection—it is there as a consequence of population dynamics.

Extinction is important in two other connections. It is fundamental to any theory of pest control, and it is in the field of economic entomology that it is possible to study real populations over long periods of time and wide areas. Therefore a satisfactory theory of extinction will make it possible to incorporate a number of applied subjects into general ecology.

Finally, we suggest that extinction itself evolves. Many of the lower invertebrates are directly limited by the physical environment so that environmental change is perhaps the major cause of their extinction, but the higher animals have much more flexible tolerances. They generally exist in a much narrower range of conditions than they could tolerate physiologically, and their physiological limits are modified in different parts of their range. Here the action of other species may be the major direct cause of extinction. Finally, some species are sufficiently dominant to modify their own environments to the point where continued survival becomes impossible.

Our strategy in investigating extinction is to work at several loosely coupled levels:

1. The general dynamics of an extinction-migration equilibrium model, in which the extinction and migration parameters are taken as given.

2. The interaction of population selection, mendelian selection within populations, selective migration, and some random phenomena in determining the changes of gene frequency in the species. Here the various parameters will be functions of gene frequency. They will be either arbitrary functions or simple models chosen to develop qualitative results. Our special concern is the outcome of selection on genes which increase the survival probability of their own population but are selected against the problem of the so-called altruistic genes.

3. The population dynamics of local populations will be examined for several different kinds of extinction in order

to develop realistic models for population parameters as functions of gene frequency.

The relations among the various processes are shown in Figure 1.

1. Demographic equilibrium in the large

Let a species' range consist of a large number of pockets of suitable habitat. Within each habitat a population is established, occasionally sends out migrants, and eventually goes extinct with some probability distribution of survival time. Thus, if $N(t)$ is the proportion of the possible sites which support populations at time t,

$$dN/dt = mg(N) - \bar{e}N \tag{1.01}$$

where $\bar{e}$ is the average extinction rate. The function $g(N)$ increases with the number of populations sending out migrants and also with the number of available sites. Therefore the essential features of $g(N)$ are preserved in the model

$$dN/dt = mN(1 - N) - \bar{e}N. \tag{1.02}$$

The substitution $y = 1/N$ thus gives the solution

$$N = N_0 \Big/ \left\{ e^{-(m-\bar{e})t} \left(1 - \frac{N_0 m}{m - \bar{e}} \right) + \frac{N_0 m}{m - \bar{e}} \right\} \tag{1.03}$$

which approaches

$$\hat{N} = 1 - \bar{e}/m. \tag{1.04}$$

The final rate of approach is

$$d(n - \hat{N})/dt = -(m - \bar{e})(N - \hat{N}).$$

Clearly a population will persist only if $\bar{e} < m$. Here we see that if the equilibrium level is near 1 a change in $\bar{e}$

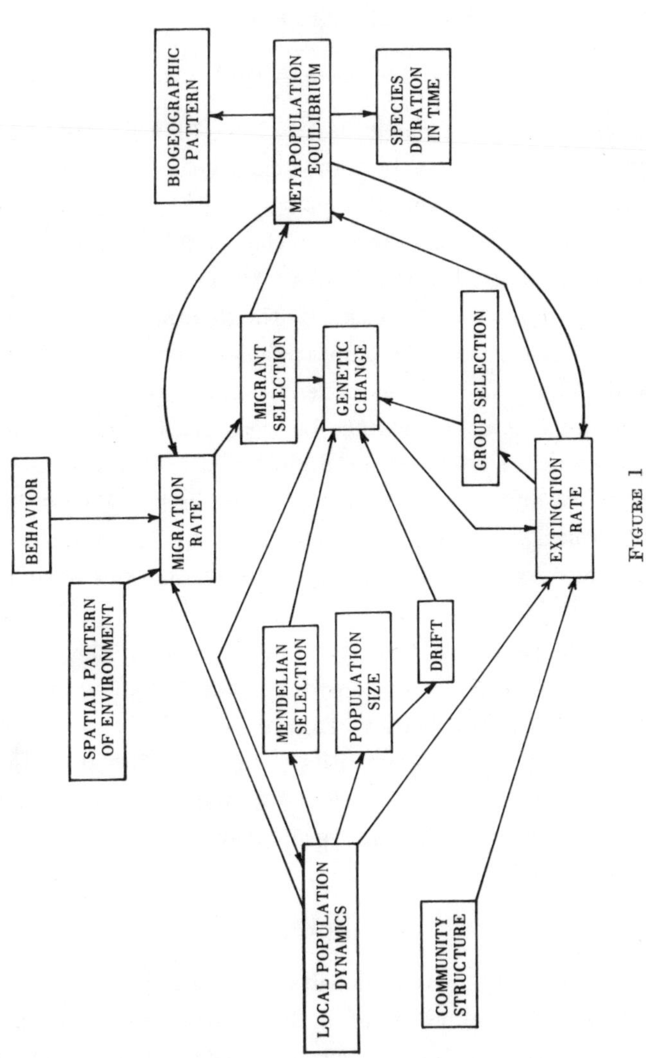
BIOGEOGRAPHIC PATTERN
METAPOPULATION EQUILIBRIUM
SPECIES DURATION IN TIME
BEHAVIOR
MIGRATION RATE
MIGRANT SELECTION
GENETIC CHANGE
GROUP SELECTION
EXTINCTION RATE
SPATIAL PATTERN OF ENVIRONMENT
MENDELIAN SELECTION
POPULATION SIZE
DRIFT
LOCAL POPULATION DYNAMICS
COMMUNITY STRUCTURE

FIGURE 1

has relatively little effect on $\hat{N}$ while if $\hat{N}$ is small it is more sensitive to changes in $\bar{e}$. Secondly, if m is large the system is very sensitive to changes in m:

$$d\hat{N}/dm = -1/m^2.$$

Therefore species which already have high migration and extinction rate may be more sensitive to selection on m and become increasingly fugitive.

The parameter m depends on the behavior of the animals and spatial distribution of sites. Although we do not know how rapidly m decreases with distance, exponential decline is a reasonable qualitative description (Mac Arthur and Wilson [**1**]). Thus an environmental change which doubles the mean distance between sites squares m and has a greater effect than a doubling of $\bar{e}$. For instance, if $m = .25$ and $\bar{e} = .1$, $\hat{N} = .6$. A doubling of $\bar{e}$ gives $\hat{N} = .2$. But a squaring of m gives $m = .06$ and the species becomes extinct. This forces us to change our notion of suitable habitat. If there is always a finite probability of extinction of a local population even in the best of circumstances, a region will be suitable or unsuitable depending on the density of appropriate sites and a species will fail to survive even if its optimal habitat is present. Along a geographic cline in the frequency of the optimal habitat, the equilibrium proportion of sites occupied will be

$$\hat{N}(D) = 1 - \bar{e}e^{a/D}, \tag{1.05}$$

where D is the relative density of good habitats.

Thus the boundary of a distribution may be quite abrupt even in the absence of marked changes in the environment.

If $m < \bar{e}$, the species moves towards extinction at a rate which approaches $(\bar{e} - m)$. However, while this is happening $\bar{e}$ may itself be decreasing due to differential survival of populations. If $\bar{e}$ falls below m, N will then increase toward some equilibrium.

The extinction rate $\bar{e}$ may also be subject to random fluctuations due to the environment. Then (1.02) becomes

$$dN/dt = mN(1 - N) - \bar{e}(t)N. \tag{1.06}$$

A formal solution of (1.06) by the substitution $N = 1/y$ gives

$$\begin{aligned} N = N_0 \exp\left(-\int \bar{e}(t)\,dt + mt\right) \Big/ \Big\{ 1 + m \\ \times \exp\left(-\int \bar{e}(t)\,dt + mt\right) \int_0^t \exp\left(\int_0^s \bar{e}(t)\,dt - mt\right) ds \Big\}. \end{aligned} \tag{1.07}$$

The probability of N falling below any given level can be found as a function of m, mean of $\bar{e}(t)$, and the variance of the $\int \bar{e}(t)\,dt$. Where the $\bar{e}(t)$ have no autocorrelation the method of diffusion equations (Kimura [**5**]) gives a limit distribution of N:

$$\varphi(N) = CN^{2((m-\bar{e})/\sigma_e{}^2 - 1)} e^{-2(m/\sigma_e{}^2)N}. \tag{1.08}$$

Here C is a constant which makes $\int \varphi(N)\,dN = 1$, $\bar{e}$ is the mean extinction rate, and σ_e^2 is its variance. Thus if $m - \bar{e} > \sigma^2$, $\varphi(N)$ rises to a single peak at $\hat{N} = 1 - \bar{e} + \sigma_e^2/m$. But if $m < \bar{e} + \sigma_e^2$, $\varphi(N)$ is a decreasing function. Most of the time N is near zero, and there is no true steady state. The species will go extinct in that region despite sporadic outbreaks.

If $\bar{e}(t)$ is autocorrelated, the diffusion equation method does not properly apply. However, an examination of

(1.07) shows that $\bar{e}(t)$ enters by way of the integral $\int_0^t e(f)\,df$, and therefore the variance σ_e^2 enters by way of the variance of the integral. A positive autocorrelation of the $\bar{e}(t)$ has the effect of increasing the variance of the integral and therefore σ_e^2 in (1.08). Hence the correlation of successive environments makes survival more difficult.

The consideration of spatial density of habitats and of temporal fluctuations of $\bar{e}(t)$ coincide in the conclusion that a species with a broad tolerance of diverse conditions has a better chance of survival, that the direction of population selection is toward increased niche breadth. However, counter pressures also exist.

2. Changes of gene frequency due to differential extinction

We now ignore the number of populations and look at the distribution of gene frequency over populations. Let $F(x)$ be the proportion of populations which have gene frequency x. Populations become extinct at a rate $e(x)$ which depends on the genetic makeup. Then after a short time interval

$$(2.01)\qquad F(t+dt, x) = \frac{F(t)[1 - e(x)\,dt]}{1 - dt \int e(x)F(x)\,dx}.$$

Therefore the change in $F(x)$ is

$$(2.02)\qquad dF/dt = -(e(x) - \bar{e})F(x),$$

where $\bar{e} = \int e(x)F(x)\,dx$. The mean extinction rate is also changing:

$$(2.03)\qquad \frac{d\bar{e}}{dt} = \frac{d}{dt}\int e(x)F(x)\,dx$$

which is

$$\frac{d\bar{e}}{dt} = \int e(x) \frac{dF}{dt} dx \tag{2.04}$$

or

$$\frac{d\bar{e}}{dt} = -\int e(x)[e(x) - \bar{e}]F(x)\, dx. \tag{2.05}$$

Thus we have the familiar result (Wright [**6**]):

$$\frac{d\bar{e}}{dt} = -\sigma_e^2. \tag{2.06}$$

Since $0 < \bar{e} < 1$, the variance of $\bar{e}$ is always less than $\bar{e}(1 - \bar{e})$. Therefore $\bar{e}$ changes toward minimum $e(x)$ at a rate slower than

$$d\bar{e}/dt = -\bar{e}. \tag{2.07}$$

Meanwhile, populations are going extinct at a rate $\bar{e}$. Therefore the species will go extinct before interpopulation selection can go to completion. We must therefore introduce colonization of new sites into the model.

3. Colonization

We can conceive of the formation of new populations in two ways. It is possible for existing populations to spread en masse, founding new populations with the same genetic make-up. Then

$$F(t + dt, x) = F(t, x) \frac{\{1 - e(x)\, dt + m\, dt\}}{1 - \bar{e}\, dt + m\, dt} \tag{3.01}$$

so that

$$dF/dt = -(e(x) - \bar{e})F. \tag{3.02}$$

Thus the migration does not enter into the dynamics of genetic change but only permits it to occur by preserving the species from rapid extinction. Now

$$(3.03) \qquad F = F_0 \exp\left(-e(x)t + \int \bar{e}(t)\, dt\right).$$

Each gene frequency increases as long as $e(x) < \bar{e}$ and then declines. Although we do not have an exact expression for the rate of change of $\bar{e}$, (3.03) gives

$$(3.04) \qquad \frac{F(x_1,t)}{F(x_2,t)} = \frac{F(x_1,0)}{F(x_2,0)} \exp\left([e(x_1) - e(x_1)]t\right).$$

This process has no real genetics. Populations either increase or decrease, but like begets like, and if a frequency was not represented originally it will not appear.

The second model of colonization assumes that a migrant pool is created from which N individuals are chosen at random to found new populations. The new populations therefore have a binomial distribution $B(x,\bar{x})$ with mean $\bar{x}$ and variance $\bar{x}(1 - \bar{x})/N$. Now we have

$$(3.05) \qquad F(t + dt, x) = \frac{F(x)\{1 - e(x)\, dt\} + mB(x,\bar{x})\, dt}{1 - \bar{e}\, dt + m\, dt}.$$

Therefore

$$(3.06) \qquad dF/dt = -(e(x) - \bar{e})F + mB(x,\bar{x}) - mF.$$

At demographic equilibrium $\bar{e} = m$ and

$$(3.07) \qquad dF/dt = -e(x)F + \bar{e}B(x,\bar{x}).$$

This has the formal equilibrium solution

$$(3.08) \qquad F(x) = \bar{e}B(x,\bar{x})/e(x)$$

where, however, $\bar{e}$ and $\bar{x}$ are functions of time. $F(x,t)$

actually approaches this limit since

$$(3.09)\quad \frac{d}{dt}\left(F(x,t) - F(x)\right) = -e(x)\left[F - \frac{\bar{e}B(x,\bar{x})}{e(x)}\right].$$

Thus the rate of approach is $e(x)$ for each x and for the whole distribution the rate of approach is faster than min (e). Meanwhile $\bar{e}$ is changing at a rate equal to the variance of $e(x)$ over populations. It will be shown that this is less than $\bar{e}$.

4. Investigation of stability properties

We will limit our consideration to the problem—does group selection ever prevail over ordinary selection, and can it result in polymorphism? In all cases we will assume that group selection alone moves the mean gene frequency toward 1, and the other directional processes move it toward 0. Therefore group selection will be said to prevail if $F(x,\bar{x})$ is stable in the neighborhood of $\bar{x} = 1$. Similarly, if $F(x,\bar{x})$ is stable near $\bar{x} = 0$, the counter selection prevails. Both may be true. Then the result of the interaction of group selection with other processes depends on the initial distribution. Finally, if $F(x,\bar{x})$ is unstable near 0 and 1, there is a permanent genetic heterogeneity. The metapopulation will be polytypic, although we do not exclude the possibility of a U-shaped distribution concentrated near 0 and 1.

Thus our method consists of deriving equations for the central moments μ_k of the distribution $F(x)$. Then the Lyapunov criterion for the stability of the distribution is that all the eigenvalues of the matrix whose elements

$$(4.01)\qquad \alpha_{ij} = \frac{d}{d\mu_j}(\mu_i')$$

have negative real parts.

This is an infinite matrix, however it has a structure imposed by the model which allows the eigenvalues to be found.

First consider a system defined by

$$(4.02) \qquad dF/dt = -e(x)F(x).$$

The function $e(x)$ can be expanded in Taylor series about the mean to give

$$(4.03) \qquad e(x) = e_0 + e_1(x - \bar{x}) + e_2(x - \bar{x})^2 \cdots.$$

Then for $\bar{x} = 0, 1, \bar{e} = e_0(\bar{x})$.

(4.04)

$$\frac{d\mu_K}{dt} = \frac{d}{dt}\int (x - \bar{x})^K \frac{dF}{dt}\,dx - K\frac{d\bar{x}}{dt}\int (x - \bar{x})^{K-1} F\,dx.$$

The second term is $-K(d\bar{x}/dt)\mu_{K-1}$. Since the matrix will be studied at an equilibrium value, $d\bar{x}/dt = 0$. And since all the central moments vanish near $\bar{x} = 0$ or 1, $\mu_{K-1} = 0$. Thus this term does not enter the matrix at all and need not be considered further. Substituting (4.02) in (4.04), the relevant part of the equation is

$$(4.04) \qquad d\mu_K/dt = -\sum e_i \mu_{K+i} \quad \text{for} \quad K > 1$$

and

$$(4.05) \quad d\bar{x}/dt = -\int (x - \bar{x} + \bar{x}) \sum e_i (x - \bar{x})^i F\,dx$$

or

$$(4.06) \qquad d\bar{x}/dt = -\sum e_i(\mu_{i+1} + \bar{x}\mu_i).$$

But

$$\sum e_i \mu_i = \bar{e}.$$

Thus

$$(4.07) \qquad d\bar{x}/dt = -\sum e_i \mu_{i+1} - \bar{x}\bar{e}.$$

The elements of the matrix can be evaluated near $\bar{x} = 0$ or 1.

$$\alpha_{11} = -\sum \mu_{i+1} \frac{d\bar{e}_i}{d\bar{x}} - \bar{e} - \bar{x}\frac{d\bar{e}}{d\bar{x}} = -\bar{e}$$

$$\alpha_{KK} = -\mu_0 e_0 = -e_0,$$

$$\alpha_{K1} = -\sum \mu_{K+1}(de_i/d\bar{x}) = 0,$$

$$\alpha_{1K} = -e_{K-1},$$

$$\alpha_{Kj} = -e_{j-K} \quad \text{for } j > K, \text{ and } 0 \text{ for } j < K.$$

Hence the matrix has all zero elements below the principle diagonal. Its eigenvalues are simply the α_{KK}.

We will show in the following sections that for the model

$$dF/dt = -e(x)F + mN(x,\bar{x}) \tag{4.08}$$

where $N(x,\bar{x})$ is a probability distribution whose only parameter is $\bar{x}$, the single element α_{21} is added to the matrix. The characteristic equation is then a product

$$P(\lambda) = \{(\lambda + \alpha_{11})(\lambda + e_0) + \alpha_{21}e_1\} \prod (\lambda + e_0). \tag{4.09}$$

The equation

$$dF/dt = -e(x)F(x) + (MF)' \tag{4.10}$$

changes the diagonal elements and introduces $\alpha_{K,K-1}$ terms which vanish. Thus its stability depends on the diagonal elements. When we generalize further to

$$dF/dt = -e(x)F(x) + (MF)' + \tfrac{1}{2}(VF)'', \tag{4.11}$$

the $\alpha_{K,K-1}$ terms do not vanish and the characteristic equation becomes a product of quadratics. Thus all these systems are analyzable in the neighborhood of the absorbing states $\bar{x} = 0$ and $\bar{x} = 1$.

In most cases we can assume that if the real parts of the eigenvalues are negative, this gives a stable distribution and positive real parts give an unstable system. However, not every eigenvalue is relevant to each moment. If there are no nonzero terms below the principal diagonal, the equations are separable into

$$d\mu_K/dt = \lambda_K \mu_K. \tag{4.12}$$

It is conceivable that the mean goes to zero but the higher moments do not. For example, some of the diffusion equations used in genetics give formal solutions

$$F(x) = Cx^{-a}(1-x)^{-b}. \tag{4.13}$$

Near 0, it can be treated roughly as Cx^{-a}. Then consider the fraction of the probability from x_0 to x_1 which lies between x_0 and x_2 $(x_2 > x_1)$:

$$P(x_0) = \frac{\int_{x_0}^{x_1} x^{-a}\,dx}{\int_{x_0}^{x_2} x^{-a}\,dx} = \frac{x_1^{1-a} - x_0^{1-a}}{x_2^{1-a} - x_0^{1-a}}. \tag{4.14}$$

Set

$$\begin{aligned} P_0 = \lim_{x_0 \to 0} P_0(x_0) &= (x_1/x_2)^{1-a} \qquad 1-a > 0, \\ &= 1 \qquad 1-a < 0. \end{aligned} \tag{4.15}$$

If $P = 1$, the distribution is mostly at 1. Nevertheless we can also calculate the ratio

$$P_1 = \lim_{x_0 \to 0} \frac{\int_{x_0}^{x_1} xF(x)\,dx}{\int_{x_0}^{x_2} xF(x)\,dx}. \tag{4.16}$$

This is 1 if $2 - a < 0$. Then we say that the mean is zero. Continuing the process we find some P_K such that all P_i, $i > K$ are less than 1. Then all moments up to K vanish but higher moments do not.

Another example of this kind of distribution is given by

$$\begin{aligned} x &= 0 \quad \text{probability} \quad 1 - 1/N^2, \\ &= \sqrt{N} \quad \text{probability} \quad 1/N^2. \end{aligned} \tag{4.17}$$

Then as $N \to \infty$, the first three moments vanish, the fourth moment goes to 1, and higher moments increase without bound.

In any case, if the mean goes to zero then x is almost always less than any ε and we will consider the distribution to go to zero.

5. Group selection vs. migrant selection

In this model migrants have local populations to form a migrant pool. But differential population sizes, survival, or propensity to migrate results in a gene pool with an altered array of gene frequencies

$$x^* = -\bar{x} - s\bar{x}(1 - \bar{x}). \tag{5.01}$$

This model assumes additive genetics. If there is dominance, the equation will depend on whether mating occurs before or after emigration. A sample of N migrants founds a new population with mean gene frequency x^* and variance $x^*(1 - x^*)/N$. We will approximate this binomial distribution by a normal distribution with the same mean and variance. Thus the distribution of gene frequencies changes according to the equation

$$dF/dt = -e(x)F(x) + \bar{e}N(x,x^*). \tag{5.02}$$

Then the central moments satisfy the equation

$$d\mu_K/dt = -\sum e_i\mu_{K+1} + \bar{e}\int (x - \bar{x})^K N(x,x^*)dx \tag{5.03}$$

and

$$d\bar{x}/dt = -\sum e_i\mu_{1+i} + \bar{e}(\bar{x} - x^*). \tag{5.04}$$

Hence

$$\alpha_{11} = -S\bar{e}(1 - 2\bar{x}),$$
$$\alpha_{KK} = -e_0,$$
$$\alpha_{12} = -e_1,$$
$$\alpha_{Kj} = 0 \quad \text{for} \quad j < K \quad \text{except when} \quad K = 1,$$

and

$$\alpha_{K1} = \bar{e}(d/d\bar{x}) \int (x - \bar{x})^K N(x,x^*)\, dx.$$

Near $x = 0$ or 1, $x^* = \bar{x}$ and the integral is the Kth moment of the distribution. Thus

$$\alpha_{K1} = \bar{e} \frac{d}{dx} (K-1)(K-3) \cdots \left(\frac{\bar{x}(1-\bar{x})}{N}\right)^{K/2} \qquad K \text{ even,} \tag{5.05}$$
$$= 0 \qquad K \text{ odd.}$$

For $K = 2$, this is $\bar{e}(1 - 2\bar{x})/N$, but for $K > 2$ it vanishes. Hence the characteristic equation

$$P(\lambda) = \prod (\lambda + e_0) \left\{ \begin{vmatrix} -s\bar{e}(1-2\bar{x}) - \lambda & -e_1 \\ \dfrac{\bar{e}(1-2\bar{x})}{N} & -e_0 - \lambda \end{vmatrix} \right\}. \tag{5.06}$$

Therefore $\lambda = -e_0$ is a multiple root, and

$$\lambda^2 + (e_0 + s\bar{e}(1 - 2\bar{x}))\lambda + \bar{e}e_1((1 - 2\bar{x}))/N + e_0\bar{e}s(1 - 2\bar{x}) = 0. \tag{5.07}$$

This gives

$$\lambda = -\tfrac{1}{2}(e_0 + s\bar{e}(1 - 2\bar{x})) \pm ((e_0 + s\bar{e}(1 - 2\bar{x}))^2 - 4(1 - 2\bar{x})\bar{e}(e_1/N + e_0 s))^{1/2}. \tag{5.08}$$

This will have negative real parts if $e_0 + \bar{e}s(1 - 2\bar{x}) > 0$ and $(e_1/N + e_0 s)\bar{e}(1 - 2\bar{x}) > 0$. The first condition is always satisfied at $\bar{x} = 0$, 1. Since $\bar{e} > 0$, the second condition is satisfied at $\bar{x} = 0$ if

$$e_1/N(0) + e_0(0)s > 0. \tag{5.09}$$

But $e_1 = e'(x)$ at $\bar{x}$. This is always negative, and measures the differential group selection. Hence we require $|e_1(0)| < Ne_0(0)s$. For the linear model $e(x) = a(1 - bx)$, stability at $\bar{x} = 0$ requires $b < Ns$, and at $\bar{x} = 1$, $b > Ns/(1 + Ns)$. Therefore if $b < Ns/(1 + Ns)$, the system is unstable at 1 and migrant selection prevails over group selection. If

$$Ns/(1 + Ns) < b < Ns,$$

the system is stable at zero and one. Either migrant or group selection prevails, depending on the initial distribution. And if $b > Ns$, group selection always prevails. The fourth possibility, a distribution which is unstable at both zero and one is excluded when $e(x)$ is linear but would occur if

$$-\frac{e'(0)}{e(0)} > Ns > \frac{Ns}{1 + Ns} > -\frac{e'(1)}{e(0)}. \tag{5.10}$$

Thus $\log e(x)$ must become flatter near $x = 1$. The derivative of the logarithm rather than $e(x)$ itself makes the condition independent of the absolute rate. This is a consequence of the metapopulation equilibrium which adjusts the extinction rate to equal the migration rate. Note also that the smallest λ is less than $\bar{e}$. Thus the mean changes more slowly than the distribution as the latter approaches $\bar{e}N(x,x^*)/e(x)$. Therefore there is a real transient limit distribution on the way to fixation.

The conditions for stability show that group selection can overwhelm migrant selection or block it from going to completion, but the interpopulation selection coefficient must be greater than the migrant selection coefficient by a factor of N, the size of new populations. Of course, if the metapopulation is decreasing, $\bar{e} > m$ and group selection can prevail more readily.

If migrant selection favored a dominant α_{11} would become $2s\bar{x}(2 - 3\bar{x})$. Then at $\bar{x} = 0$ the equation becomes

$$(5.11) \qquad \prod (\lambda + e_0)\left\{\begin{vmatrix} 0 - \lambda & -e_1 \\ \bar{e}(1 - 2\bar{x})/N & -e_0 - \lambda \end{vmatrix}\right\}.$$

Then the quadratic is

$$(5.12) \qquad \lambda^2 + e_0\lambda + e_1\bar{e}/N = 0.$$

Thus the system is unstable at $\bar{x} = 0$. Near $\bar{x} = 1$, the quadratic is

$$(5.13) \qquad (\lambda - s\bar{e})(\lambda + e_0) - \bar{e}e_1/N = 0.$$

This is stable if $-e'(i)/e(1) > Ns$.

This is the same condition as before, since near $\bar{x} = 1$ there are no homozygous dominants. But near $\bar{x} = 0$ there are mostly dominants and heterozygotes and therefore virtually no migrant selection. If migrant selection favored a recessive the condition near $\bar{x} = 0$ would be unchanged, but near 1 the equation becomes

$$(5.14) \qquad \lambda^2 + e_0\lambda = e_1\bar{e}/N = 0.$$

Since $e_1 < 0$, the system is always stable. Near $\bar{x} = 1$, there are only dominant homozygotes and heterozygotes and hence virtually no migrant selection. Hence migrant selection for a dominant cannot block group selection. There is either permanent polymorphism or fixation at $\bar{x} = 1$. Migrant selection for a recessive results in fixation at $\bar{x} = 1$ if $-e'(0)/e(0) > Ns$, and fixation at either end depending on starting conditions if $-e'(0)/e(0) < Ns$.

We expect that intermediate levels of dominance will behave qualitatively like additive selection. If migrants first mate within local populations, the genotype frequencies depend on F. The frequency of homozygotes will be a function of σ_x^2 as well as $\bar{x}$, and the results will resemble the additive case.

6. Group selection vs. mendelian selection

Consider now the model of within-population mendelian selection at a rate $M(x)$ reducing gene frequency x toward 0, and group selection increasing x as before.

The probability that a population with gene frequency x is transformed into one with gene frequency $x - dx$ in time dt is $dtM(x)$. Then, under this process alone,

$$(6.01)\quad dF/dt = -M(x)F(x) + M(x + dx)F(x + dx).$$

Expanding the right side,

$$(6.02)\qquad dF/dt = (MF)',$$

for $0 < x < 1$, and 0 at 0, 1.

Since for selection $M(0) = 0$, the matrix has no non-zero elements below the principle diagonal. The stability of the process therefore depends on the diagonal elements. The equations for the moments are now

$$(6.03)\qquad \frac{d\mu_K}{dt} = \int (x - \bar{x})^K (MF)'\, dx.$$

Integrating by parts thus gives

$$(6.04)\quad \frac{d\mu_K}{dt} = (x - \bar{x})^K MF\Big|_0^1 - K\int (x - \bar{x})^K MF\, dx\,.$$

For $K = 0$, $d\mu_0/dt = 0$ so that $MF|_0^1 = 0$. For distributions near $\bar{x} = 0$ or 1, $(x - \bar{x})^K MF|_0^1 = 0$ as well, and this term can be dropped.

Expanding $M(x)$ in Taylor series about the mean, we get

$$\frac{d\mu_K}{dt} = -K \sum_{i=0}^{\infty} M_i \mu_{K+i-1} \tag{6.05}$$

or

$$\frac{d\mu_K}{dt} = -K \sum_{K-1}^{\infty} M_{j-K+1} \mu_j. \tag{6.06}$$

Therefore $\alpha_{KK} = -KM_1$.

$$\frac{d\bar{x}}{dt} = \int x(MF)'\,dx \tag{6.07}$$

which is $-\sum M_i \mu_i$. Thus

$$\alpha_{11} = -dM_0/d\bar{x}.$$

But this is $-M_1$ at $x = \bar{x}$. Thus for additive selection

$$M = (s\bar{x}(1 - \bar{x} + s(1 - 2\bar{x})(x - \bar{x}) - s(x - \bar{x})^2)) \tag{6.08}$$

and $M_1 = s(1 - 2x)$. The diagonal element is $-KM_1$. Thus the system is stable at $\bar{x} = 0$, unstable at $\bar{x} = 1$. Mendelian selection prevails over group selection. If mendelian selection acts against a recessive $M_1 = s\bar{x}(2 - 3\bar{x})$. This is passively stable at $\bar{x} = 0$ and unstable at $\bar{x} = 1$. But with mendelian selection favoring a recessive, $M_1 = s(1 - \bar{x})(3\bar{x} - 1)$. This is now passively stable at $\bar{x} = 1$ and unstable at $\bar{x} = 0$.

Thus group selection is swamped completely by mendelian selection except when mendelian selection favors a total recessive.

The interesting interactions between mendelian selection and group selections arise when some random process creates new interpopulation variance. This can occur either through the founding of new populations or by drift within populations.

7. Group selection vs. mendelian selection with small founding populations

Now the full equation for change is

$$dF/dt = -e(x)F(x) + \bar{e}N(x,\bar{x}) + (MF)'. \tag{7.01}$$

Thus the equations for the moments are

$$\frac{d\mu_K}{dt} = -\sum e_i\mu_{K+i} + \bar{e}\theta_K - K\sum_{K-1}^{\infty} M_{i-K+1}\mu_i. \tag{7.02}$$

Here θ_K is the Kth moment of the normal distribution with mean $\bar{x}$ and variance $\bar{x}(1-\bar{x})/N$, and

$$d\bar{x}/dt = -\sum e_i(\mu_{1+i} + \bar{x}\mu_i) + \bar{e}\bar{x} - \sum M_i\mu_i. \tag{7.03}$$

This is

$$d\bar{x}/dt = -\sum e_i\mu_{1+i} - \sum M_i\mu_i. \tag{7.04}$$

The diagonal elements are

$$\alpha_{11} = -M_1, \qquad \alpha_{KK} = -e_0 - KM_1.$$

$$\alpha_{12} = -e_1 - M_2, \qquad \alpha_{21} = \bar{e}(1-2\bar{x})/N$$

and all other α_{Kj}, $j < K = 0$. Hence the characteristic equation is

$$\prod_{K=3}^{\infty} (\lambda + e_0 + KM_1) \times \begin{bmatrix} -M_1 - \lambda & -e_1 - M_2 \\ \bar{e}(1-2\bar{x})/N & -e_0 - 2M_1 - \lambda \end{bmatrix} = 0. \tag{7.05}$$

The infinite product gives all negative eigenvalues if $M_1 \geq 0$. Therefore for additive selection the system is always unstable at $\bar{x} = 1$. Stability at $\bar{x} = 0$ depends on the quadratic

$$(M_1 + \lambda)(e_0 + 2M_1 + \lambda) + (e_1 + M_2)[(1-2\bar{x})\bar{e}]/N = 0. \tag{7.06}$$

For additive selection

$$M(x) = s\bar{x}(1 - \bar{x}) + s(1 - 2\bar{x})(x - \bar{x}) - s(x - \bar{x})^2. \tag{7.07}$$

Thus at $\bar{x} = 0$ we get

$$(s + \lambda)(e_0 + 2s + \lambda) + (e_1 - s)\bar{e}/N = 0. \tag{7.08}$$

Thus

$$\lambda^2 + (e_0 + 3s)\lambda + s(e_0 + 2s) + \bar{e}(e_1 - s)/N = 0. \tag{7.09}$$

Therefore

$$\lambda = -\tfrac{1}{2}\{e_0 + 3s \pm ((e_0 + 3s)^2 - 4[s(e_0 + 2s) + \bar{e}(e_1 - s)/N])^{1/2}\}. \tag{7.10}$$

Thus the system is stable near $\bar{x} = 0$ if

$$Ns(e_0 + 2s) + \bar{e}(e_1 - s) > 0, \tag{7.11}$$

or

$$-e'(0) < (N - 1)s + 2Ns^2/e(0). \tag{7.12}$$

In order to compare this result with § 5, recall that the intensity of migrant selection was $\bar{e}s$, and the condition for stability was $-e'(0)/\bar{e} < N\bar{e}s/\bar{e}$. Dividing (7.12) by $\bar{e} = e_0$,

$$-e'(0)/e(0) < (N - 1)[s/e(0)] + 2N(s/e(0))^2. \tag{7.13}$$

For $N = 4$ and

$$s/e(0) = .1, \qquad -e'(0)/e(0) < 3.8 \tag{7.14}$$

compared to .4 in § 5.

In general if $e(0)$ is large enough the requirement for stability at $\bar{x} = 0$ is more stringent than in the migrant selection case. The reason for this is that for migrant selection the genetic variance is $\bar{x}(1 - \bar{x})$ but for mendelian selection it is $\bar{x}(1 - \bar{x}) - \sigma_x^2$. However, (7.13) is not linear in $s/e(0)$. An increase in the extinction rate reduces the chance for stability at $\bar{x} = 0$.

Thus group selection cannot prevail completely over mendelian selection but can produce polytypy if it is stronger than the mendelian selection. Thus as we go from the center to the periphery of a species' distribution where extinction rates are greater we should reach a threshold beyond which group selection counters mendelian selection and we will have marginal polymorphism.

For selection against a dominant,

$$\begin{aligned} M(x) &= s\bar{x}(1 - \bar{x})^2 + s(1 - \bar{x})(1 - 3\bar{x})(x - \bar{x})4 \\ &\quad - s(2 - 3\bar{x})(x - \bar{x})^2 + s(x - \bar{x})^2. \end{aligned} \tag{7.15}$$

Near $\bar{x} = 1$, $M_1 = 0$ and $M_2 = s$. Thus the characteristic equation is

$$\lambda(e_0 + \lambda) - \frac{\bar{e}}{N}(e_1 + s) = 0. \tag{7.16}$$

Then we have stability if $e_1 + s < 0$ or $-e'(1) > s$. Hence it is much easier for group selection to fix a gene whose mendelian effects are dominant. Near $\bar{x} = 0$, $M_1 = s$ and $M_2 = -2s$, the characteristic equation is then

$$(\lambda + s)(\lambda + e_0 + 2s) + (\bar{e}/N)(e_1 - 2s) = 0. \tag{7.17}$$

Thus we require

$$-e'(0) < (N - 2)s + 2Ns^2/e_0. \tag{7.18}$$

This condition is only slightly more stringent than the additive case. Therefore if

$$-e' > (N - 2)s + 2Ns^2/e_0$$

group selection prevails over mendelian selection. If

$$s < -e' < (N - 2)s + 2Ns^2/e_0$$

the result is fixation either at 0 or 1 depending on the initial distribution, and if $-e' < s$ mendelian selection prevails.

8. Group selection vs. mendelian selection plus drift within populations

When there is sampling variance within populations the equation for gene frequency change becomes (see Kimura [5])

$$dF/dt = -(e(x) - \bar{e})F + (MF)' + \tfrac{1}{2}(VF)''. \tag{8.01}$$

$M(x)$ is the mean directional change and $V(x)$ the variance, $= x(1 - x)/N$ where N is the size of local populations. Then

$$\begin{aligned} d\mu_K/dt = {} & -\sum e_i\mu_{K+1} + \bar{e}\mu_K - K\sum m_{i-K+1}\mu_i \\ & + \frac{K(K-1)}{2}\sum V_{i-K+2}\mu_i \end{aligned} \tag{8.02}$$

and

$$d\bar{x}/dt = -\sum e_i\mu_{i+1} - \sum M_i\mu_i. \tag{8.03}$$

Therefore

$$\alpha_{11} = -M_1,$$

$$\alpha_{KK} = -e_0 + \bar{e} - KM_1 + \frac{K(K-1)}{2}V_2.$$

But $\bar{e} = e(0) = e_0$, and

$$\alpha_{KK} = -KM_1 + \frac{K(K-1)}{2}V_2.$$

With additive selection,

$$M_1 = s(1 - 2\bar{x}), \qquad V_1 = \frac{(1 - 2\bar{x})}{N},$$

$$M_2 = -s, \qquad V_2 = -1/n.$$

Thus

$$\alpha_{11} = -s(1 - 2\bar{x}),$$

$$\alpha_{KK} = -Ks(1 - 2\bar{x})4 - \frac{K(K-1)}{2n},$$

$$\alpha_{12} = -e_1 + s,$$

$$\alpha_{K,K-1} = \frac{K(K-1)}{2} V_1 = \frac{K(K-1)}{2n}(1 - 2\bar{x}),$$

$$\alpha_{K-1,K} = -e_1 + (K-1)s.$$

There are no other α_{Ki} below the principal diagonal. Thus the matrix can be expanded into a product of quadratics of which the first is

$$(8.04) \quad (\lambda + s(1 - 2\bar{x}))(\lambda + 2s(1 - 2\bar{x}) + 1/n) + (e_1 - s)[(1 - 2\bar{x}]/n) = 0$$

and the subsequent ones are

$$(8.05) \quad \left(\lambda + (K-1)s + \frac{(K-1)(K-2)}{2n}\right) \times \left(\lambda + ks + \frac{K(K-1)}{2n}\right) + (e_1 - (K-1)s) \times \frac{K(K-1)}{2n}(1 - 2\bar{x}) = 0.$$

Near $\bar{x} = 0$ the first equation becomes

$$(8.06) \quad (\lambda + s)(\lambda + 2s + 1/n) + (e_1 - s)/n = 0.$$

Then stability requires that

$$(8.07) \quad s(2ns + 1) + e_1 - s > 0$$

or $-e_1 = -e'(0) < 2ns^2$. Equation (8.05) gives the requirement

$$K(K-1)\left[s + \frac{K-2}{2n}s + \frac{K-1}{2n}\right] + [e_1 - (K-1)s]\frac{K(K-1)}{2n} > 0, \tag{8.08}$$

or

$$(2ns + K - 2)(2ns + K - 1) + e_1 - (K-1)s > 0.$$

This is

$$4n^2s^2 + 2ns(2K-3) - (K-1)s + (K-1)(K-2) > -e'(0).$$

It is always satisfied. Therefore mendelian selection prevails over group selection only if $-e'(0) < 2ns^2$. For $s = .1$, $n = 10$, this requires $-e'(0) < 2s$.

Near $\bar{x} = 1$, (8.04) becomes

$$(\lambda - s)(\lambda - 2s + 1/n) = (e_1 - s)/n = 0. \tag{8.09}$$

Now stability requires

$$-s(1/n - 2s) - (e_1 - s)/n > 0, \tag{8.10}$$

or

$$2ns^2 - s + s - e_1 > 0.$$

Therefore the system is stable at 1 if $-e'(1) > 2ns^2$. Hence if $-e'(x) > 2ns^2$ at $x = 0$ or 1, group selection prevails over mendelian selection. If

$$-e'(0) < 2ns^2 < -e'(1),$$

the initial distribution determines the outcome. If

$$-e'(0) > 2ns^2 > -e'(1),$$

there will be permanent polytypy, and if $e'(x) < 2ns^2$ mendelian selection will predominate. The effective population size n in a population which fluctuates is close to its minimum value. In our case this will be near the average number of offspring of the first colonist.

We can now compare the criteria for group selection blocking mendelian selection when the interpopulation variances comes from drift within populations and when it is derived from colonization. In the previous section, the conditions were more stringent because the rate of creation of variance is tied to the migration (and hence extinction) rate. Despite the fact that the number of colonists is less than the average population size, a population is only founded once whereas intrapopulation drift occurs in each generation.

9. Group selection vs. mendelian selection with variable selection coefficients

Random variation in the intensity and direction of selection can result in the creation of interpopulation variance and therefore might drive group selection. The equations are the same as in the model of the previous section except that

$$V(x) = \sigma_s^2 x^2 (1 - x)^2. \tag{9.01}$$

Thus

$$V_0 = \sigma^2 \bar{x}^2 (1 - \bar{x})^2, \qquad V_1 = 2\sigma^2 \bar{x}(1 - \bar{x})(1 - 2\bar{x})$$

and

$$V_2 = \sigma^2[(1 - 2\bar{x})^2 - 2\bar{x}(1 - \bar{x})].$$

The diagonal elements of the matrix are

$$\alpha_{11} = -M_1,$$

$$\alpha_{KK} = -KM_1 + \frac{K(K-1)}{2} V_2 = -KM_1 + \frac{K(K-1)\sigma^2}{2},$$

$$\alpha_{K,K-1} = -KM_0 + \frac{K(K-1)}{2} V_1 = 0$$

and that

$$\alpha_{21} = dV_0/d\bar{x} = 0.$$

Thus this matrix has the eigenvalues $\lambda_K = \alpha_{KK}$. Since

$$K(K - 1)/2$$

increases more rapidly than K, α_{KK} is eventually positive and the system is unstable independently of any group selection. Thus environmental fluctuations acting on additive mendelian selection are insufficient to drive group selection to completion.

Although the distribution departs from the limit $\mu_K = 0, \bar{x} = 0$, each eigenvalue corresponds to a single moment. Although the higher moments increase, $\bar{x}$ still returns to zero and σ_x^2 may do so if $\bar{s} > \sigma_s^2$. However the peculiarities of this system are irrelevant to the problem of group selection.

10. Applicability

We have shown that although group selection is less effective than migrant or mendelian selection, it is capable of predominating over migrant selection and at least establishing polymorphism and polytypy against mendelian selection. The question arises, under what circumstances can the migrant or mendelian selection be in the opposite direction from group selection?

In the case of migrant selection there is less difficulty since the migrants are in quite a different environment from the established populations. For instance, it seems as if winged aphids are less resistant to bad weather and are less fertile than wingless aphids.

The interaction of traits affecting mendelian and group selection is more complex.

1. Selection within populations tends to increase r, the intrinsic rate of increase, but this increases the oscillations of the populations. Since the probability of extinction is a roughly exponential function of current size, increased oscillation increases the probability of random extinction.

2. A large population size may increase the probability of using up the food resource, may attract predators sooner, and may contaminate the environment more rapidly than it can be removed. Since selection within stable populations increases the carrying capacity K (and usually the population size), there will be an antagonism between group and mendelian selection here.

3. Many studies on the competition among genotypes show that there is facilitation. That is, the total population size is greater in a mixed population. But it still may be that at any mixture level one type is favored. Then mendelian selection goes in one direction and group selection toward an intermediate frequency. The stability properties near $\bar{x} = 0$ are still formally like those for linear extinction.

4. Many adaptive traits impose a cost on their bearers. In the absence of the danger (infection, predation, etc.) those who lack these traits are favored. But this increases the danger. Thus mendelian selection goes toward some level which is frequently dependent while group selection is unidirectional.

Summary

The metapopulation concept is introduced as a population of populations which go extinct locally and recolonize. A region is suitable for a metapopulation if the mean extinction rate is less than the migration rate. Thus a species can survive even if it does not form a part of any stable local community. The nature of the species' boundary is discussed in terms of metapopulation stability.

Local extinction can result in group selection if the extinction rate is a function of gene frequency. Group selection favoring one homozygote results in a transient limit distribution on the way to fixation.

Group selection may be opposed by selection among migrants. This will be a consequence of behavioral differences in the propensity to migrate, differences in population size producing a migrant pool with a gene frequency different from the mean of the population means, or differential viability.

Group selection depends on the interpopulation variance in gene frequency. But this variance is being exhausted both by group selection and by other kinds of selection. Therefore group selection can determine the outcome of opposing selections only if new variance is being created by the sampling to found new populations or drift within populations. Random variation in intrapopulation selection coefficients does not help group selection.

The stability of limit distributions due to the interaction of group and migrant or intrapopulation selection was analyzed through an infinite set of simultaneous equations in the moments.[1] It was found that group selection can prevail over migrant selection, but only if it is more intense. Stable polymorphisms are also possible and in some cases the initial distribution of gene frequencies determines the final outcome. Group selection is

[1] The structure of the resulting matrices was used to find the eigenvalues.

far more effective when the mendelian or migrant selection favors a recessive. Drift within populations is more likely to result in group selection than in random colonization. Group selection is especially likely to be important at the ecological margin of a species' distribution.

REFERENCES

1. R. H. Mac Arthur and E. O. Wilson, *The theory of island biogeography*, Princeton Univ. Press, Princeton, N.J., 1967.

2. G. C. Williams, *Adaptation and natural selection*, Princeton Univ. Press, Princeton, N.J., 1966.

3. E. O. Wilson and D. S. Simberloff, *Experimental zoogeography of islands*. I, II, Ecology **50** (1969) 278–314.

4. V. C. Wynne-Edwards, *Animal dispersion in relation to social behavior*, Hafner, New York, 1962.

5. M. Kimura, *Diffusion models in population genetics*, J. Appl. Probability **1** (1964), 177–232. MR **30** #2946.

6. S. Wright, *Classification of factors of evolution*, Cold Spring Harbor Symposium on Quantitative Biology, Vol. 20, 1955, 16–20.

THE TEMPORAL MORPHOLOGY OF A BIOLOGICAL CLOCK

By

ARTHUR T. WINFREE

Princeton University

"If, however, the anatomist only recognizes his scheme and overlooks the individual reality of his objects, then he is a Castalian and a beadplayer who is using mathematics on a most unsuitable object."

Hermann Hesse
Magister Ludi

1. Introduction

An approximately diurnal periodicity has been found in the behavior or physiology of almost every major variety of living organism. Though in some instances these circadian rhythms may be driven by environmental periodicities, in others they are apparently of autonomous origin. We inquire into the ultimate source of this innate rhythmicity, the basic "driving" oscillator or "clock". Very little is known of its concrete (presumably biochemical) mechanism. Even within the province of formal dynamics, several fundamental questions have remained unanswered, e.g.:

1. Is there just one, or more than one, important state variable or degree of freedom?
2. Does the clock have an equilibrium or stationary state (a singularity in the dynamical flow)?
3. Is there one or more than one singularity?
4. Are these singularities stable or not?
5. Is the clock self-exciting?
6. Is the clock a limit-cycle oscillator? A relaxation oscillator? A damped oscillator?

This report shows one approach to such inquiries, using phase resetting experiments. The original purpose of these experiments was to *prove* that the Drosophila clock is essentially a very stable self-exciting limit-cycle oscillator. This seemed an overwhelmingly plausible inference from the success of the Pittendrigh-Pavlidis-Ottesen entrainment model, in which [**1**], [**2**], [**3**] the effect of perturbation is inferred to consist primarily of an instantaneous phase shift: immediately after perturbation the oscillator appears to resume its preferred mode of oscillation, with all properties normal, except for a predictable advance or delay of phase.

In order to exclude alternatives to the limit-cycle hypothesis, a perturbation method was developed for the experimental investigation of complex nonlinear oscillatory processes. This method is based on the fortunate circumstance that for most circadian rhythms there exists a kind of perturbation which will reset the phase of the rhythm without permanently affecting the period. In this case our perturbation is a short single pulse of visible light, which resets the daily time of emergence of flies from their pupal cases. Our most important observable is therefore the centroid time (the arithmetic mean) of the daily emergence peak of a population of flies, measured from the end of the light pulse. This interval, which we will call *cophase*, θ, is measured modulo the steady-state period (approximately 24 hours), which is taken as the unit of time.[1]

A discussion of facts and inferences about the fruitfly system, which are important for the rigorous interpretation of the experiments sketched here, may be found in the papers of Pittendrigh [**7**] and in my dissertation, in preparation [**4**]. But the purpose of this summary is not

[1] Qualitatively, cophase measures the interval from the end of the perturbation to the periodic occurrence of the clock-regulated event. It therefore measures something about the "state" of the clock just after perturbation. Our cophase, θ, is the *complement* of the quantity $\mathcal{O}$ introduced by Ottesen to represent the "new phase" to which the clock is immediately reset by a light pulse. In a limit-cycle process, cophase is the complement of the "latent phase" [**5**] associated with the state reached by the perturbation.

so much to support particular hypotheses about the Drosophila clock as to illustrate a method of inquiry into the topological dynamics of oscillatory control processes. We attempt to design simple, crucial experiments utilizing θ measurements in order to exclude broad classes of oscillator models as suggested above. The discriminating criteria are based on topological more than quantitative features of the dependence of θ on two parameters of the perturbation: S, its duration, and T, the interval between initiation of the oscillation and application of the perturbation.

2. Perturbation experiments: format and logical constraints

We consider first the basic design of a phase-resetting experiment. Continuous white light (LL) inhibits the rhythm of emergences. We induce emergence activity rhythms in replicate aliquots of pupae by transfer to darkness (DD) after prolonged exposure to LL, as in [**7**].

Figure 1 illustrates an idealized phase-resetting experiment. Rhythmicity is initiated in 20 replicate cultures of pupae (vertical tracks) of mixed developmental ages by LL/DD transfer at different times (along the 45-degree line). The circles represent emergence centroids, following LL/DD at (nominal) 24-hour intervals in each culture. When we simultaneously illuminate all the pupae for a duration S, the various cultures are hit at all times T after initiation.

At this point an interval of transient irregularity ensues, not shown in Figure 1. It is followed by reappearance of 24-hour rhythmicity, but with reset phase. Following Pittendrigh, we call this post-transient interval the "steady-state". It is here that we observe the dependence of emergence time (plotting the daily centroids as circles downward) upon T (plotted to the right; we show an experiment every three hours along the T axis). We will call this function the "resetting map" since it maps the "old phase" of the rhythm just prior to perturbation into the "new phase" just after perturbation. Figures 1(a) and (b) show two kinds of resetting maps which may result. The interval from perturbation to the emergence centroids, ignoring multiples of 24 hours, will be called the "cophase" θ.

Let us assume that resetting maps are *continuous* and make the approximations

(1) that the overt rhythm responds identically to stimuli given at time T or at time $T \pm 24$; and

(2) that the overt rhythm returns to its former period after this kind of perturbation.

Then it follows that the average (over a cycle of T) slope of a resetting map can take on only integer values.

That is, insofar as these idealizations are realistic, Experiment E of Figure 1 is *identical* to Experiment E′: in both cases S seconds of light are given at the same stage of the cycle. So the two sequences of subsequent emergence centroids must be identical. That is, the resetting map must have wriggled vertically through zero or an integer number of cycles as we moved laterally from E to E′ through one cycle of T. If the resetting is a continuous curve, its average slope must therefore be an integer, and the steady-state data of a graph such as Figure 1 will resemble a wallpaper.

Figure 1 illustrates two possible cases: in (a) the perturbation had little effect, so the resetting map almost

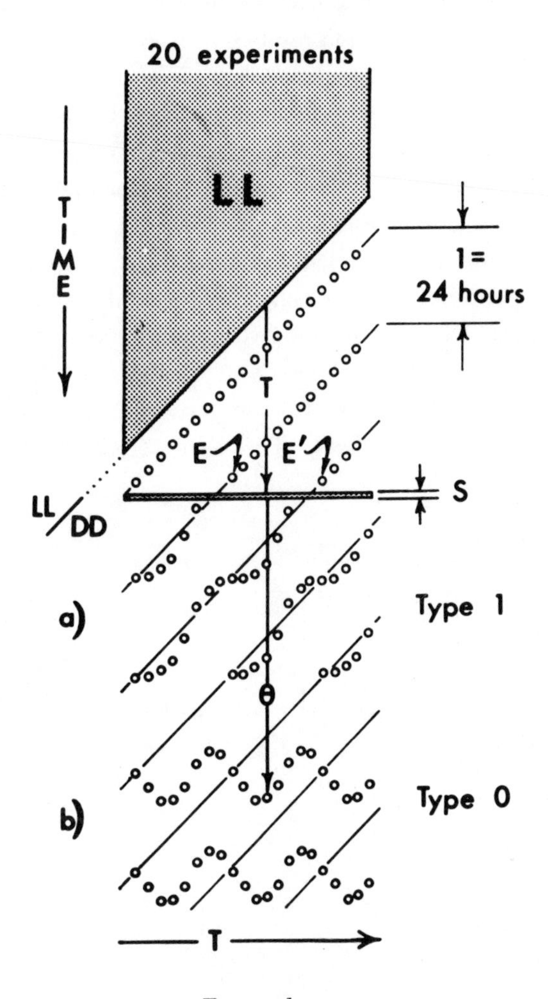

FIGURE 1

Idealized format for measuring resetting maps and the cophase surface $\theta(T,S)$. The circles in the 20 vertical tracks represent centroids of daily emergence peaks in 20 independent and differently-perturbed populations of fruitfly pupae.

parallels the controls, with average slope, or "Type", $= 1$; in (b) the perturbation resets the rhythm to a phase which is relatively independent of T, with average slope, or "Type", $= 0$. Theoretically admissable but not illustrated are types with average slopes $+2$, -1, etc.

Now consider a whole series of resetting maps for a given circadian system and kind of perturbation: one for each duration S. We stack the resetting maps one behind the other in order of increasing S as a third coordinate axis. Cophase is now seen as a two-dimensional surface $\theta(T,S)$ over the stimulus plane, $T \times S$. We will draw strong inferences about the dynamics of the underlying clock process from the shape of this surface (or of its contour map; Figure 2).

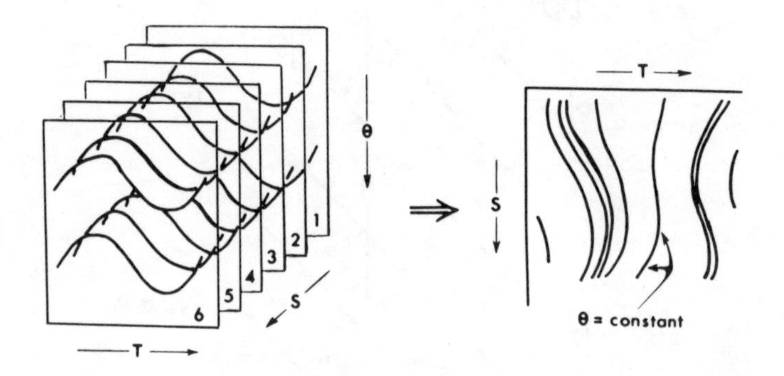

FIGURE 2
The cophase surface $\theta(T,S)$ is a composite of resetting maps $\theta(T)$ generated as S varies (LEFT). It may be represented as a contour map showing the loci of fixed θ on the $T \times S$ plane (RIGHT).

3. Two empirical generalizations

First we make use of available facts. Perturbation experiments of the sort diagrammed in Figure 1 have long been used to measure the phase shifts of the physiological and behavioral rhythms of diverse plant, animal, and unicellular organisms. In fifteen published cases (see Table 1) the resetting map is given or can be adequately reconstructed from given data. Two empirical generalizations emerge:

(A) In all cases the resetting map is either Type 0 or Type 1; no cases with average slope -1, $+2$, etc. are reported.

(B) In several cases two different types are obtained with the same organism by simply changing the duration of the perturbation: the Type 0 always corresponds to the more prolonged illumination.

4. Cophase contours: by induction

The above generalizations are based on a wide variety of circadian rhythms. But perhaps the same kind of clock process underlies them all. If so, we might try to use generalizations (A) and (B) as clues to the major qualitative features of the cophase surface $\theta(T,S)$ of that kind of process. We will try to sketch the contours of constant cophase on the stimulus plane, $T \times S$.

As S is increased, Type 1 resetting maps are replaced by Type 0, and no further types are known. We conjecture therefore that for any circadian system and perturbation, there will be a critical S^* separating Type 1 resetting maps (for "weak" or short perturbation) from Type 0 (for "strong" or prolonged perturbation).

The resetting maps are *sections* through $\theta(T,S)$ at fixed S. Below S^*, therefore, we encounter every θ contour along any line parallel to the $S = 0$ axis (see Figure 1a). All cophase contours begin on that axis, since every cophase is reached during an unperturbed free-run. But at $S > S^*$, those cophases which can be achieved are achieved at *two* T's (see Figure 1b): the cophase contours are evidently turning back toward $S = S^*$. In fact we know they have all turned back before S becomes very large, for prolonged illumination brings the clock to a fixed cophase independent of its history.

So we are led to postulate the existence of a place on the $S = S^*$ line where all the cophase contours are confluent. That is, we are led to the curious conjecture that there exists a critical stimulus, (T^*,S^*), which does *not* reset the rhythm, but *abolishes* it, for the "resetting" is evidently to all phases or no phase (Figure 3).

5. The "pinwheel" experiment

At this point it becomes interesting to retire to the laboratory and actually *map* $\theta(T,S)$ using the convenient light-sensitive clock which regulates termination of metamorphosis in the fruitfly. We do this in principle by the following single experiment, though due to technical constraints it was actually done piecemeal:

We cover a square grid with pupae in constant light, and then initiate clock oscillations in parallel vertical strips by slowly covering the grid with an opaque square, moving right to left in 24 hours. At this point all are in the dark, but those further to the right have been in the dark longer. We have thus established the T axis horizontally. Now by sliding the square cover *vertically* to expose the whole grid by the end of a three minute interval, and instantly replacing it, we give perturbations of 0–3 minutes along a vertical S axis.

After a few days transients will have subsided, and waves of emergence activity will be seen sweeping daily across

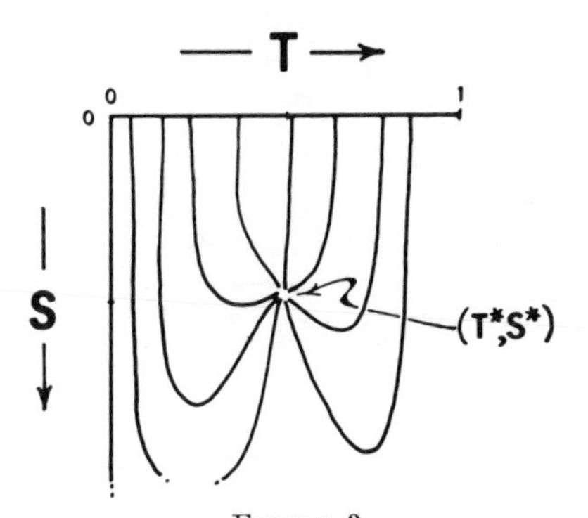

FIGURE 3
Contour map of the cophase surface $\theta(T,S)$, inferred from two empirical generalizations about the resetting behavior of circadian rhythms.

the grid. At any instant the position of the wave marks a locus of constant cophase, $\theta(T,S)$ = constant. It turns out, as expected from Figure 3, that these waves do not sweep exactly *across* the grid, but rather counterclockwise *around* a fixed center inside the grid.

In actual practice this experiment was executed in 179 separate parts, each sampling θ at one point in the grid: in each a small culture is taken from LL into DD, and at hour T given S seconds of light. Emergence events are automatically recorded, and the centroids of daily peaks are plotted as dots at a distance θ (= hours from perturbation) from the $T \times S$ plane.

In Figure 4 we plot $\theta(T, S)$ in two orthogonal projections. On the left, the projection shows points at $T < T^*$ as larger circles and $T > T^*$ as smaller. T from 15 to 24 hrs is deleted, as this portion of the surface forms a tilted plane (visible in the other projection) which would obscure the spiral ramp between $T = 0$ and $T = 15$ hrs. The right-hand projection shows data at $S = 0$ to 30 as dots, 31 to 60 as tiny circles, 61 to 90 as larger ones, and the

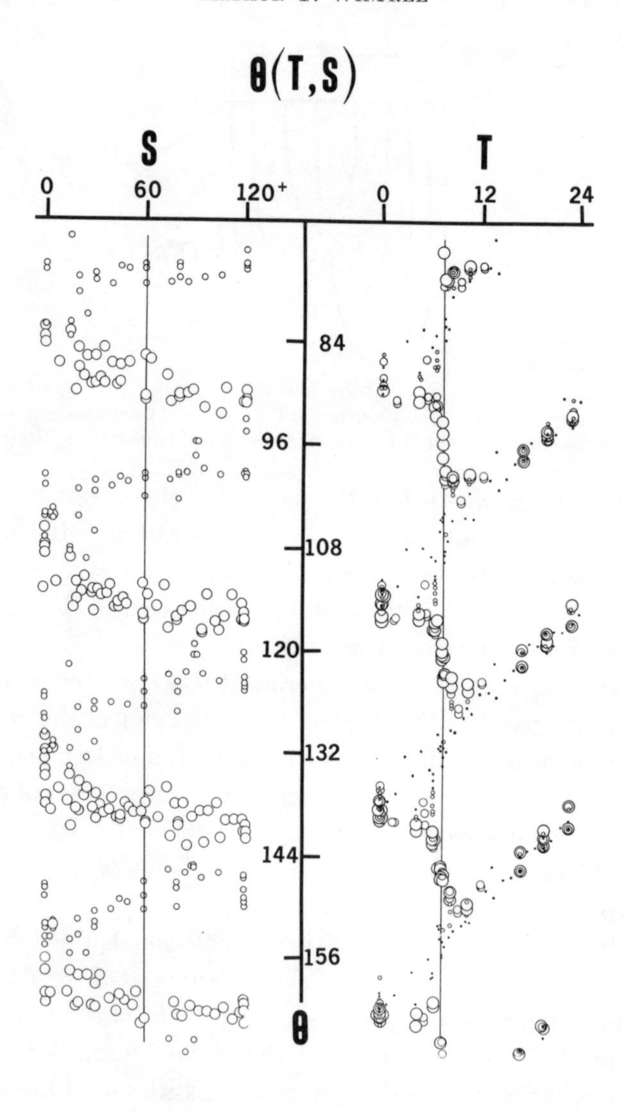

FIGURE 4
The cophase surface $\theta(T,S)$, measured using the fruitfly's pupal emergence "clock". Larger circles represent nearer data points in the two orthogonal projections.

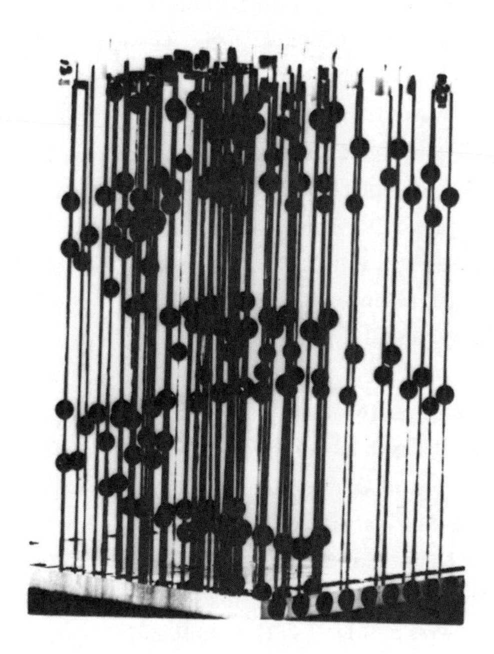

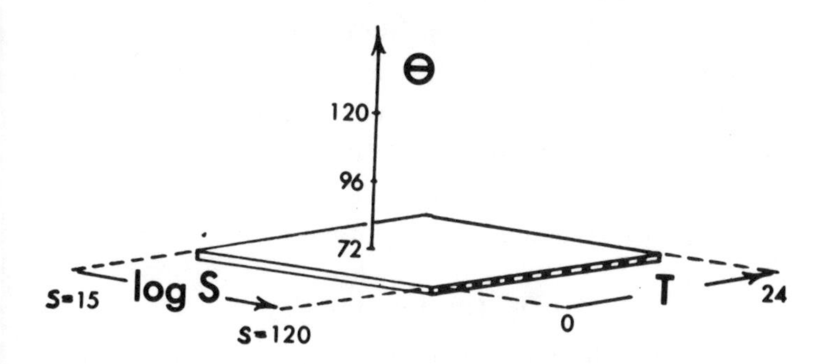

FIGURE 5
The cophase surface graphed as a wire and lucite model. The line drawing indicates co-ordinate axes in the photograph.

largest circles represent emergences following perturbations longer than 90 seconds. The data wind up around the ($T = 7.3$, $S = 60$ secs) axis, throwing off a planar region which connects with the helicoid in the next cycle of T (not shown).

Figure 5 is a photograph of about $\frac{1}{3}$ of the data, plotted as described except that (1) the S scale is logarithmic, and (2) the cophase axis is vertically *upward* so that the vertical wires will not fall out of their sockets. The lucite baseplate represents $\theta = 72$ hours after perturbation. About $2\frac{1}{2}$ cycles of emergence centroids are shown, up to $\theta = 132$.

The dots seem to outline a stack of tilted planes linked by a vertical right-handed corkscrew. The corresponding contour map and emergence waves may be visualized by considering the sections of this surface at successively higher vertical planes. They resemble Figure 3: the anticipated radial convergence of contour lines near (T^*, S^*) corresponds to the helicoidal center of our surface, which blends into the tilted planes represented by the peripheral regions of roughly parallel contour lines.

Whereas sections of $\theta(T,S)$ at fixed θ are emergence waves, sections at fixed S are resetting maps: below S^* (which turns out to be about 60 seconds), they are all Type 1, whereas above S^* they are Type 0, consistent with the empirical generalizations from more diversified data.

6. Implications

What does this tell us about the dynamical nature of the underlying clock process?

1. First of all, from the convergence of the cophase contours, i.e., from the existence of a critical stimulus

which sends the clock to "no phase or all phases", one can show that there is a *state* of the clock process to which no phase can be assigned. This appears to exclude models in which the clock is thought of as being always at some phase-point in its "cycle".

2. Secondly, the existence of a region of $\theta(T,S)$ where $\partial\theta/\partial T$ is positive (Figure 6) tells us that the state of the clock is varying in at least two independent ways during the perturbation. This can be shown by the following argument: Suppose that the clock's state were given by some *single* variable or measurement, e.g., phase, ϕ. Then in Figure 6b the ϕ axis represents all possible states of the oscillator in their cyclic order. By definition the "state" determines its own rate of change at each instant in a given

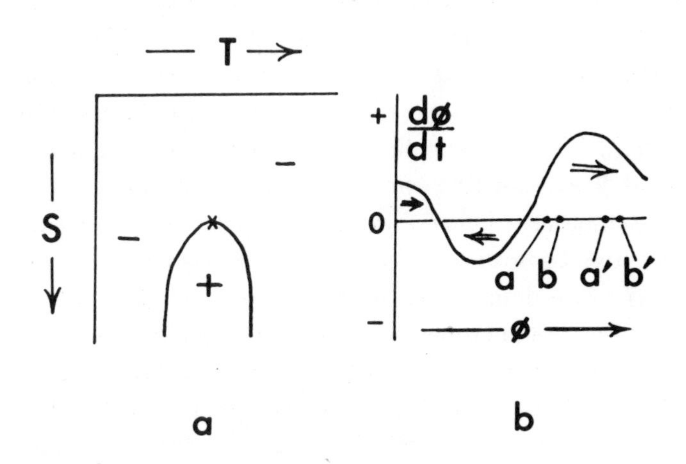

FIGURE 6

(a) The region of the cophase surface in which $\partial\theta/\partial T$ is positive. (This is the region of *negative* slope, with θ plotted downward.)

(b) The rate of change of the internal state variable, "phase" as a function of phase for a hypothetical "simple clock" in constant light.

environment. Therefore $d\phi/dt$ is one function of ϕ under standard conditions (always positive) and another during perturbation. We apply a perturbation during which ϕ changes continuously from a to a', as in Figure 6b. The curve in Figure 6 represents the phase velocity, $d\phi/dt$ associated with each phase, ϕ. Oscillator B was initiated infinitesimally earlier than A so that at the time of perturbation oscillator B is at a slightly later phase than A. During the perturbation oscillator A cannot pass B because this would imply that two oscillators in the same environment can temporarily have the same phase but differ in phase velocity, contrary to the assumption that phase uniquely defines the oscillator's state. Therefore at the end of the perturbation the phase of oscillator B must be slightly greater than the phase of oscillator A: $b' > a'$ in Figure 6b. Now the cophase of B is slightly less than the cophase of A, i.e., $\theta_B < \theta_A$, but T_B was greater than T_A, i.e., $\partial\theta/\partial T$ is less than 0. So our observation that $\partial\theta/\partial T$ *can be* positive shows that the oscillator's state *cannot* be uniquely determined by phase alone. There must be some second state variable.

By another argument from the topology of the measured cophase surface, it can be shown that this second degree of freedom cannot be attributed merely to an effect of light on the photoreceptor: to an increasing adaptation during the interval S. The clock is not just advancing or delaying in phase during the perturbation (as a relaxation oscillator or any other "simple clock" [**18**] does), but some other aspect of its state is simultaneously being altered.

By experimental test, not only the phase, but also the shape of the resetting map for later perturbations is found to be altered by single perturbations. In particular, there exist *non*-phase-shifting combinations (T,S) which markedly reduce the amplitude of the subsequent resetting map, even to zero [**4**], [**17**].

Beyond these two conclusions, we will consider the question with all possible abstractness, subject to the constraints of available resetting data, in an attempt to derive not necessarily the simplest, but rather the *most general* model consistent with present knowledge. We can then prune the tree of alternatives by means of crucial exclusion experiments, whereas by taking a less abstract viewpoint, we might obtain a simple and plausible model, but would not know which of its features are indispensible, and should one of its implications be contradicted by experiment, we would have to start over with a new arbitrary model.

Our approach is related to those of Fitzhugh [**12**] and Pavlidis [**2**], [**8**].

7. Geometrical dynamics

First of all except at (T^*,S^*), no discontinuities appear in the graph of $\theta(T,S)$, so we infer that the clock's dynamics is continuous. Secondly we assume that the clock's state at any instant is definable by a finite number of measurements $(N \geq 2)$. So the internal state of the clock process can be put in 1:1 correspondence with the points in a N-dimensional Euclidean coordinate space. Its dynamical behavior in standard conditions is given by the spontaneous rate of change of each of the N coordinate variables. Thus at each point of the state space there is a little arrow, defined by these N rates of change, and the field of arrows defines a flow along continuous trajectories. This field of flow is a pictorial summary of

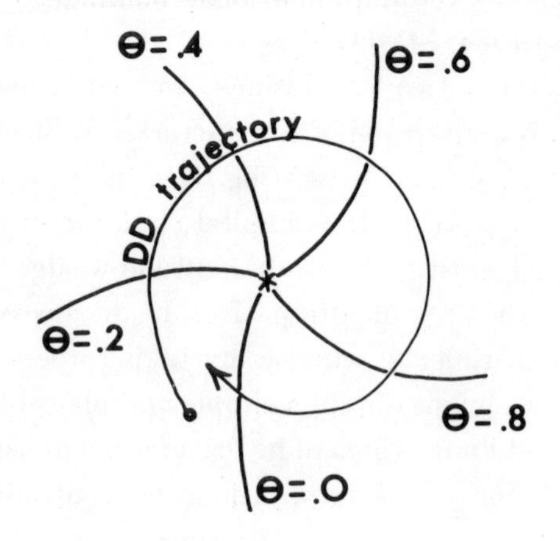

FIGURE 7
One cycle of dark dynamics, starting at the state reached in prolonged LL. Five contours of the cophase field are shown converging to the singularity in this dynamical space.

the temporal behavior of the clock in standard conditions. In altered conditions, e.g., during illumination, the behavioral flow is somehow different.

This geometrical language will prove most helpful, for it permits us to bring to bear upon subtle questions one of our most powerful hereditary tools, viz., geometrical intuition. Our arguments derive entirely from visualizing the geometry, and are purely conjectural for N greater than 3.

Interest centers on what we can observe, viz., cophase: to each possible state of the clock at time $(T + S)$ (just after a perturbation), there corresponds a unique emergence time (mod 1, time unit of 24 hrs) in the steady-state several cycles later. Intuition prompts us to regard the set of states which all have the same cophase (that is, those states which result in emergence at the same time, modulo 1) as $N - 1$ dimensional manifolds. We will call these ISOCHRONS (Figure 7).

8. Unperturbed clock behavior

Since under standard conditions we are dealing with an oscillatory process [**7**], we know that the dynamical trajectories form more or less closed loops "around" (in N-dimensional space) a central stationary state or singularity.[2] Whether these trajectories wind into the singularity as a final resting state (damped oscillation) or only circle it in closed loops (conservative oscillation) or spiral out to a limit-cycle or even to a relaxation-oscillation or show even more complex behavior, we do not know at this point.

But we do know that in every cycle of oscillation, every value of cophase is passed through once. Each isochron must cut across each loop of every trajectory surrounding the singularity. States reached along any trajectory at intervals of an integer number of time units all lie on the same isochron.

9. "Phaseless" states

So the isochrons, which cannot intersect, must be confluent in a manifold of at least $N - 2$ dimensions, thus stratifying the state space in a pinwheel-like fashion. The states constituting this manifold seem to have all cophases or no cophase . . . curiously reminiscent of the perturbation (T^*, S^*). This manifold is in some sense

[2] Perhaps more accurately described as a *singular point*, since only the scalar isochron field suffers a singularity at this state; the dynamical vector field only experiences a zero.

"phaseless", for trajectories of the unperturbed dynamics do not enter or leave it. A perturbation moving the state of the oscillator into this manifold will apparently abolish the familiar oscillation. In the case $N = 2$ this is clear, for the phaseless manifold consists entirely of a single state, the singularity, which is not only "phaseless", but also a completely motionless state of the system.

10. Clock behavior during perturbation

Now what about clock behavior in the presence of our perturbing agent? We do not consider separately the problem of stimulus transduction: whether only the on-switch or only the off-switch is effective, or the photoreceptor adapts, etc. We incorporate any such behavior as additional degrees of freedom of the comprehensive "clock process", and seek to describe the flow in this N-dimensional space, simply hoping that $N = 2$ or 3 will suffice.

We know this behavior is not oscillatory, for in continued light (LL) the rhythmicity of emergence dies out with the time-constant of transients. Moreover, after the light has been on for *any* time greater than twelve hours, a transfer back to DD results (after transients) in the same pattern of emergence [**16**] as though the system had been waiting at a stable sink, and resumed motion from there along a standard dark trajectory. (Indeed, this is how we routinely start an experiment, initiating identical oscillations in batches of pupae by putting them into the dark after sustained light.)

Even as little as a minute of dim light profoundly alters the state of the clock, so in the absence of further details, let us conjecture that LL dynamics consists of smooth,

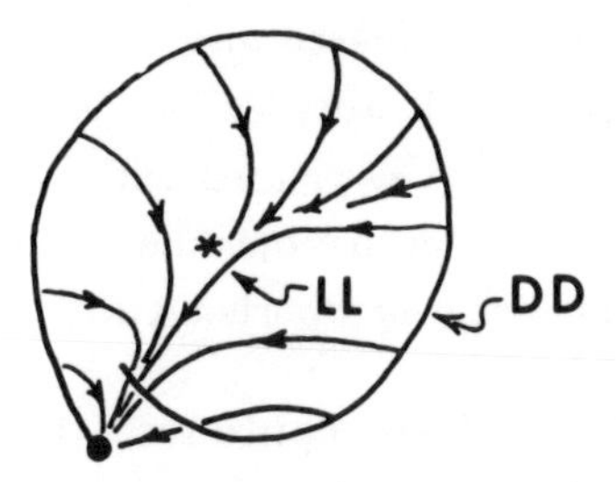

FIGURE 8
The general features of dynamical flow under constant illumination. Only the trajectories leading *from* states in the first cycle of dark dynamics are shown.

initially-rapid decay toward a stable sink, which is the point of origin of trajectories after the transition from LL to DD, as in Figure 8.

It should be noted that "DD" and "LL" refer to two points on a *continuum* of possible light intensities. For intermediate intensities, the appearance of the dynamical flow is presumably intermediate between Figures 7 and 8.

11. The significance of (T,S) in the dynamical space

So it appears that we can reach *any* state in the surface composed of LL trajectories starting on the DD trajectory which starts at the LL sink, by the simple expedient of releasing the system into a dark free-run for T hours, then switching to LL dynamics for the interval needed to push the clock any given distance towards the LL sink (Figure 8). This is exactly what we have done in the conventional resetting experiment. This procedure, in fact, *maps* this area from the dynamical space into the $T \times S$ plane.

Since the $N - 2$ dimensional "phaseless" locus at the confluence of isochrons passes *through* the DD loop trajectory, it must intersect any 2-dimensional surface bounded by that trajectory, which is to say that by measuring the cophase associated with each state in that surface, we should be able to project an image of the intersection and its immediate neighborhood into the $T \times S$ plane.

12. The cophase contours: theoretical

What will the image $\theta(T,S)$ look like? From some point on the $T \times S$ plane the isochrons, or constant-θ loci, will fan out radially; at large S we will be close to a single value of θ, regardless of initial T; at small S, we are following the DD loop trajectory, and so will cut all isochrons in sequence, cyclically. In short, this interpretation of clock dynamics gives just the same strange shape of $\theta(T,S)$ as was anticipated by induction from the two empirical generalizations about Type 1 and Type 0

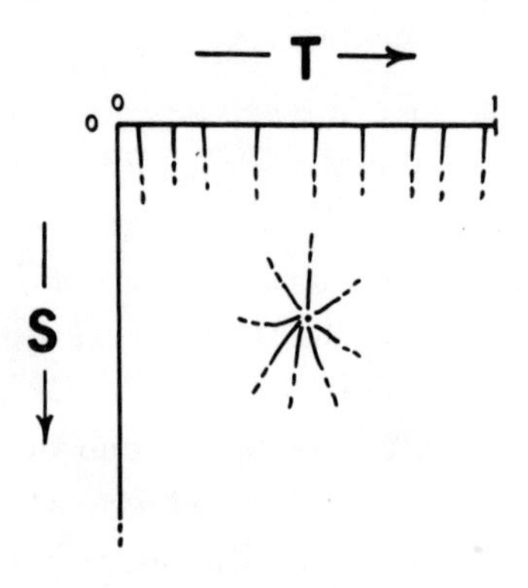

FIGURE 9
Pieces of a contour map of $\theta(T,S)$, derived as implications of the sketches in Figures 7 and 8.

resetting maps (Compare Figures 9 and 3).[3] After both these lines of inference indicated the experiment would be worthwhile, this $\theta(T,S)$ was actually measured on the circadian system in Drosophila (Figure 4).

The agreement of prediction and fact may indicate that we are on the track of a valid interpretation of clock dynamics. If so, then certain further implausible predictions must be confirmed, and a number of further experiments are required to reduce somewhat its generality.

13. The singularity: criterion one

For example: (T^*,S^*) is the symmetry axis of the cophase helicoid, and the point of confluence of the isochron contours. If this perturbation sends the clock to a "phaseless" state in the sense that it lies on a trajectory which remains in the phaseless manifold, and does not return to normal oscillation, then we expect the rhythmicity of emergences to decay with the time-constant of transients (in the order of days). This first test is not strong: the transients cannot be expected to vanish *altogether* during the remaining several days of metamorphosis, and there are other possible explanations for an apparent loss of rhythmicity. In Figure 10 the lower panel shows 4 days of emergence activity starting 4 days after perturbations not close to (T^*,S^*). The very bottom record is an unperturbed control, $(T,0)$. The upper panel, by contrast, shows the results of 5 independent experiments in which perturbations approximating (T^*,S^*) were used ($T = 6.5$ to 7.7 hrs, $S = 45$ to 65 secs).

[3] A simplified two-dimensional model described by flow similar to Figures 7 and 8 gives $\tan \theta(T, S) = -\sin 2\pi T/(S + \cos 2\pi T)$. The contour map of this surface consists of arcs of sinusoids converging at $(T^*,S^*) = (\frac{1}{2}, 1)$ as in Figure 3.

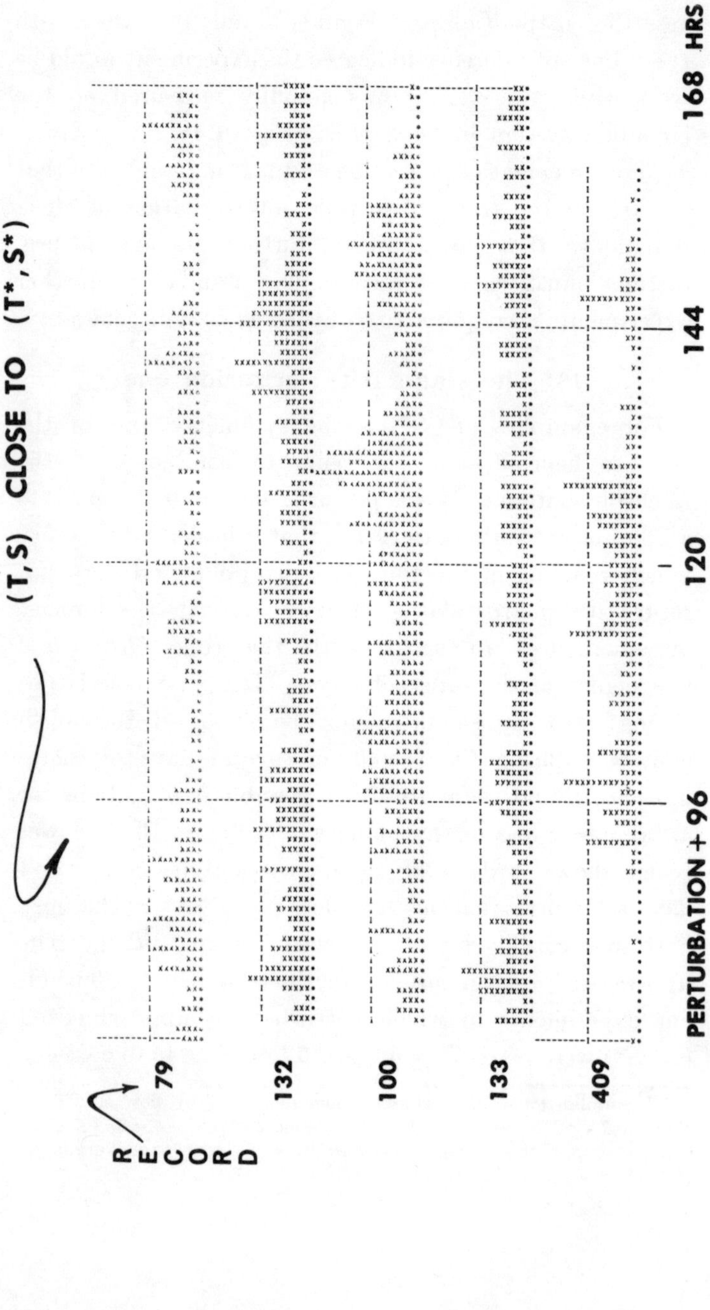

(T,S) CLOSE TO (T*,S*)

RECORD

79

132

100

133

409

PERTURBATION + 96

120

144

168 HRS

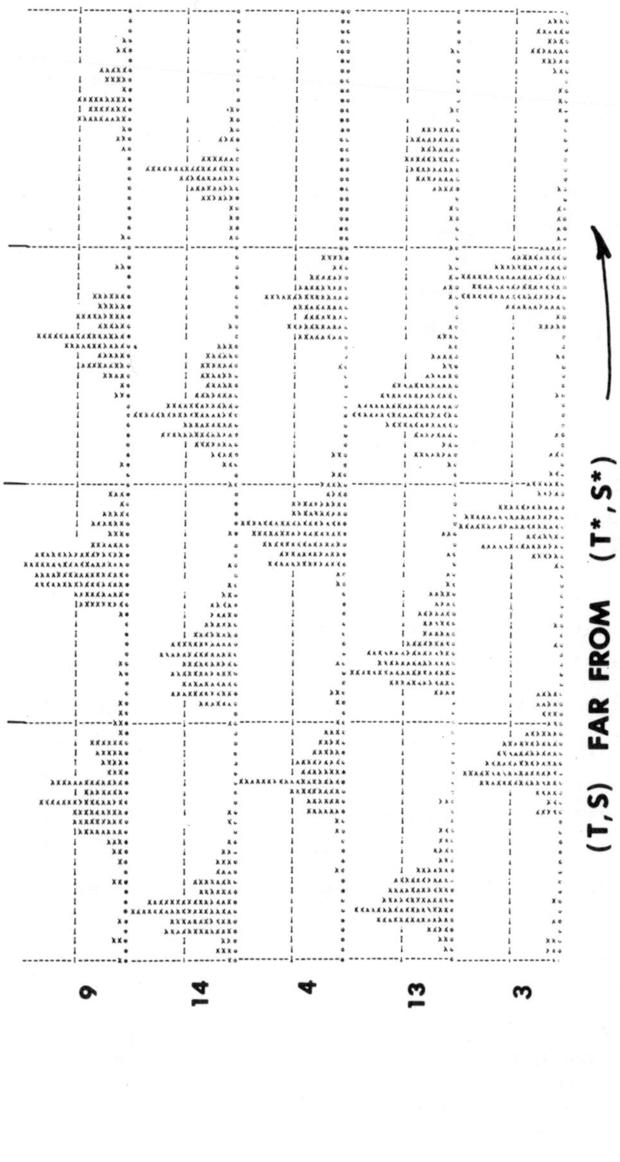

FIGURE 10

Sequences of hourly fruitfly emergence counts, starting four days after perturbations which place the photosensitive oscillator close to its stationary state (upper panel) and (lower panel) not particularly close to it. The upper panel seems nearly arrhythmic, the lower normally rhythmic. Each x = one fly.

Figure 11 gives a more comprehensive survey of the 179 experiments with which we sampled the ideal "pinwheel experiment". R is a measure of arrhythmicity which could rise as high as 140 to 150 if Poisson-distributed emergences occurred hour after hour at a constant average rate [**4**]. R is plotted above the stimulus plane, $T \times S$. Two orthogonal projections are shown. It seems clear that we approach total arrhythmicity near (T^*, S^*), and only there. But the data are quite variable.

14. The singularity: criterion two

More critically, we expect that the rhythmicity of *responsiveness* to subsequent perturbation will immediately and completely vanish, i.e., the resetting map will become a flat horizontal line. A lasting distortion of the resetting map had never been observed in any circadian system, as far as I know; a fortiori, annihilation of its periodicity by an ephemeral, weak perturbation would be a new phenomenon.

From the arguments above, we expect T^* to lie near that T at which 12 hour phase shifts can be obtained, and S^* to be smaller than any S which measures a Type 0 resetting map. So the acceptibility of this view of the clock process hangs on the paradoxical prediction that the oscillation will be completely abolished by a *delicate* tap at just the right moment, even though a more powerful perturbation, or the same perturbation applied at a different time, merely resets the clock without causing this catastrophe.[4]

[4] The probable existence of an annihilating pulse is implicit in several published clock models (e.g., Kalmus), and was independently recognized explicitly by both Pavlidis [**8**] and Michael Rosenzweig (personal communication). Engelmann [**19**] obtained a similar phenomenon *experimentally*, in Kalanchoe, using *two* pulses in a different theoretical context.

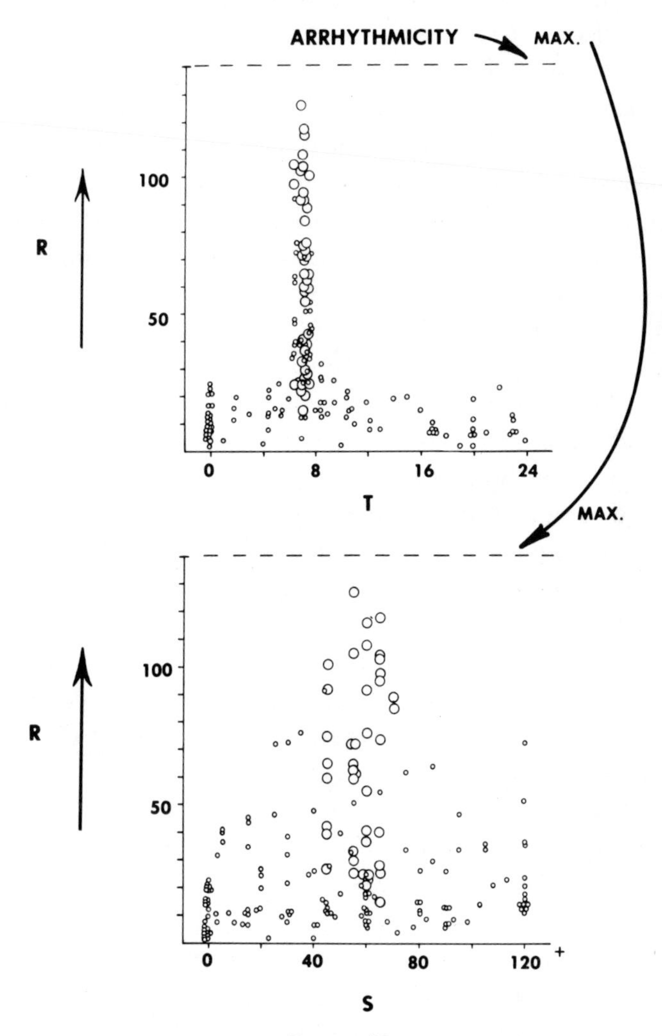

FIGURE 11

A measure of arrhythmicity, applied to the 179 perturbation experiments sampling the $T \times S$ plane. Utterly arrhythmic, Poisson distributed emergences give $R = 140$–150 in populations of a few hundred flies. R has a peak where θ has its symmetry axis and nonregular point. The larger circles in these two orthogonal projections represent perturbations particularly close to (7, 60).

The necessary experiment is to induce clock oscillations by LL to DD transitions, wait until T^* (6.5–7.5 hrs) and give S^* (45–65 sec) of standard light. In four replicate aliquots we also give 120 second "assaying pulses" 6, 12, 18, or 24 hours after the annihilating pulse. If, despite the outward appearance of arrhythmicity following (T^*, S^*), the underlying photosensitive oscillator was only phase shifted, then the assaying pulses will outline the familiar large amplitude resetting map, $\theta(T, 120)$, somehow phase shifted. On the other hand, if (T^*, S^*) put the clock into a phaseless manifold, then the time of emergences measured from the assay pulses will not show a 24-hour periodic dependence on perturbation time.

In fact there is no such dependence, as Figure 12 shows: the time from assay pulse to emergence centroid is the same ($\pm$ typical noise level) in all four cases of the three independent trials of this experiment.

An attempt to fit a low-amplitude sinusoidal resetting map to these data suggests that the amplitude is nearly zero: the phases of the "best" sinusoids differ randomly on the three panels; the amplitudes are 1.8, 2.1, 2.4 hours peak-to-peak (cf. normal 7.0); the data vary randomly ± 1.0, 0.9, 1.0 hours about these "best" sinusoids. These data seem better described as varying ± 1.2, 1.3, 1.2 hours about a perfectly flat resetting map at $\theta = 118 \pm 24n$ hours. This noise level is normal for our phase measurements: unperturbed control centroids vary ± 1.5 hours.

This is a very interesting result, for only in the case $N = 2$ is it a priori necessary that the clock's responsiveness to subsequent perturbation be not only arrhythmic, but also time-independent, as it appears to be. Only in the

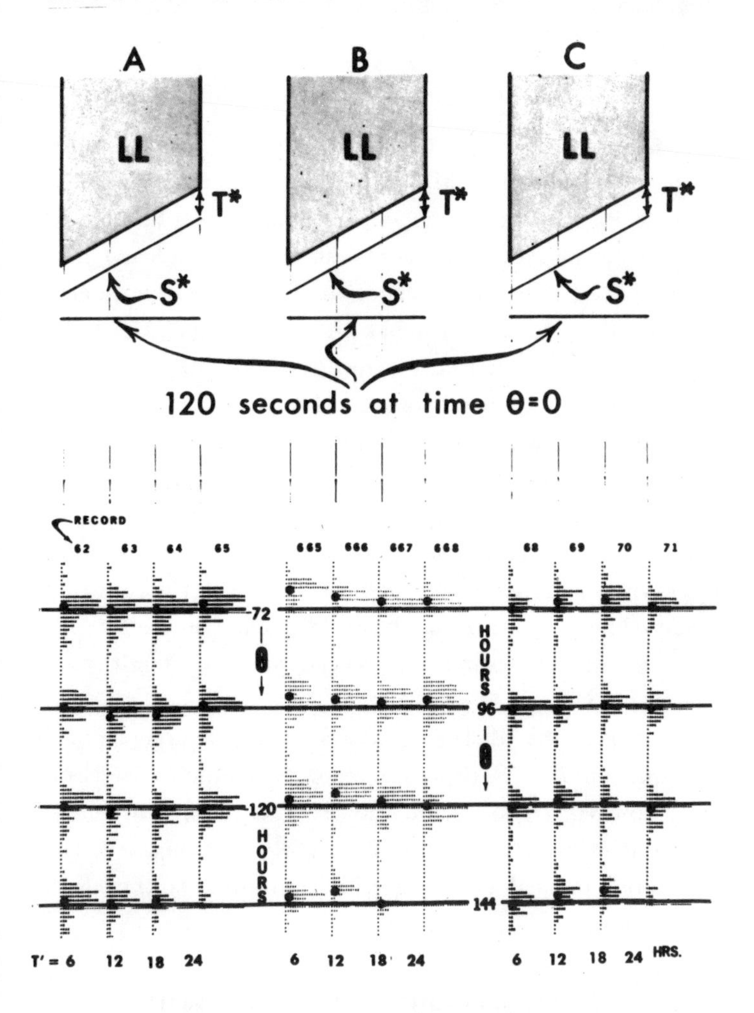

FIGURE 12
Three independent trials of an "annihilation and reinitiation" experiment, or measurement of the resetting map after application of (T^*,S^*). T' is the interval between (T^*,S^*) and the 120-second assaying perturbation at time $\theta = 0$. Again, each x = one fly emergence.

case $N = 2$ is the singularity the *only* state in the $N - 2$ dimensional phaseless manifold.

15. The number of clock variables

If the clock has only two important variables of state, then our 2-dimensional illustrations are more realistic than we had reason to hope. We should now be able to predict (or *post*dict) the constant time (modulo 24 hrs) from assay pulse to emergence in the above experiment. For if (T^*, S^*) really reaches the singularity, then the clock process is essentially "frozen" there until the assay pulse pushes it further along the LL trajectory from T^*, thus reinitiating the oscillation when it swings onto a DD trajectory. That is, an assay pulse of duration S should have the same resetting effect, *whenever* given, as does $(T^*, S^* + S)$. The latter was measured in our 1-pulse experiments mapping the cophase contours: the resulting event time is marked by horizontal bars on Figure 12, and agrees well with the observed phase of reinitiation after (T^*, S^*).

It appears by all three criteria—the convergence of θ contours on $T \times S$, the suppression of emergence rhythmicity toward (T^*, S^*), and the "correct" constant phase of reinitiation—that 7 hrs, 60 sec is, or is close to, the predicted critical stimulus which annihilates the clock oscillation.

16. The stability of the singularity

However there is an alternative interpretation. We are operating on a *population* of noninteracting oscillators (pupal circadian clocks). They are not perfectly identical, nor are they perfectly identically illuminated. Both the overt arrhythmicity of emergence and the time-independence of responsiveness could be due *not* to annihilation of the oscillations at a stable singularity, but to random scattering of phases near a very *un*stable singularity, such as would be expected in a relaxation oscillator (if it could be perturbed towards its singularity) or a strongly self-exciting limit-cycle oscillator. These classes of oscillators constitute the most plausible and most frequently proposed models for the phenomenology of entrainment by approximately diurnal light cycles.

Calling it Hypothesis B, let us consider the possibility that we are dealing with a population of self-exciting oscillators which (T^*,S^*) reduced to near-zero amplitude in each pupa, and that these regenerative oscillators quickly spiralled back out to the former amplitude, and are now all running normally but incoherently on the limit-cycle, in contrast to what we tacitly assumed earlier, that after (T^*,S^*) the oscillators *remain* near a stable singularity, i.e., near zero amplitude.

If so, then the effect of the assay pulses in Figure 12 was not reinitiation, but resynchronization. We can use the measured resetting map of the assaying pulses to predict the distribution of cophases *following* the assay pulse. For pulses with $S > S^*$, it turns out to be conspicuously bimodal, which is why pulses of 120 sec were chosen. The distribution of emergences is this bimodal distribution of phases *convoluted* with the normal distribution of emergences corresponding to a perfectly coherent population [**4**].

Figure 13 shows not the theoretical prediction, but the actual control result of executing a resynchronization experiment (simulating Hypothesis B) on an artificially constructed incoherent population of normally-running clocks, and numerically inverting the convolution [**4**].

By contrast, the same procedure applied to the "reinitiation" experiment of Figure 12 produces the unmistakably unimodal distribution of phases in Figure 14 . . . which is the expectation if the assay pulse actually initiates oscillations from a relatively stable state. This appears to exclude Hypothesis B.

A further check is obtained by examining in the same way the well-known arrhythmicity of populations reared as larvae in continuous darkness at constant temperature. If as in Hypothesis B the singularity is unstable, i.e., the clock oscillator is self-exciting and therefore (if $N = 2$) of relaxation or limit-cycle character, then such populations are equivalent to the artificially constructed population of asynchronously running clocks. Yet a 120 sec pulse produces the *unimodal* distribution of phases in Figure 14,

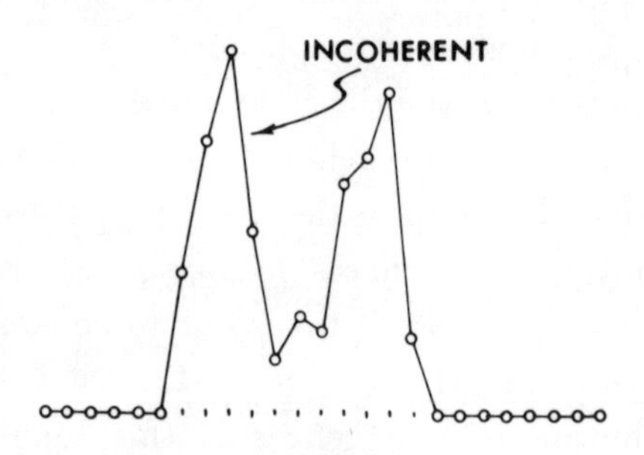

FIGURE 13
The distribution of cophases in a partially-resynchronized population of clocks, constructed by mixing subpopulations given LL/DD initiations at 2 hour intervals during one full day. This bimodal distribution was computed from the emergence distribution by numerically inverting the convolution, using as basis the typical emergence peak of a coherent population.

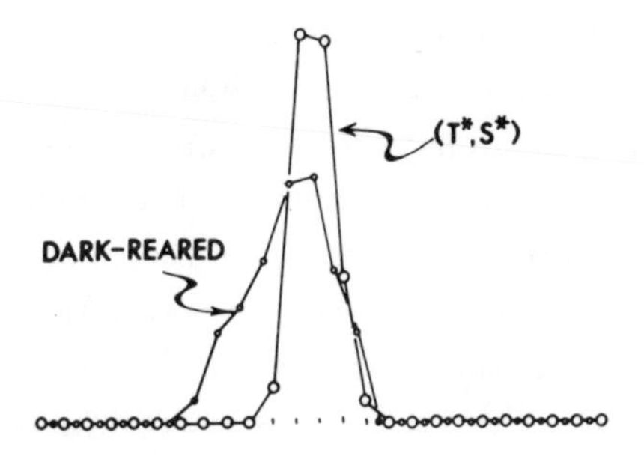

FIGURE 14
Unimodal cophase distributions, computed from emergence peaks following application of the same light pulse as in Figure 13 to populations whose relative arrhythmicity is due to (1) dark-rearing, and (2) annihilation by (T^*,S^*).

centered near the same phase as in the reinitiation after (T^*,S^*). Identical processing of the published data of Honegger [**9**] on pulse perturbation of dark-reared populations leads to the same conclusion: Hypothesis B appears to be excluded.

By a different line of reasoning and an elegantly simple experiment with a different strain of D. pseudoobscura, Zimmerman [**10**] has also concluded that temperature-shock induces rhythmicity in dark-reared populations by initiation rather than by synchronization.[5]

---

[5] In both Zimmerman's and my own experiments, there remain unexcluded (but perhaps implausible) alternatives, e.g., maybe annihilated and dark-reared populations consist of clocks running incoherently and normally *except that* their sensitivity to light and temperature shocks is somehow greatly increased.

17. The classical model

In 1964 the basis for a very successful empirical model of entrainment behavior was provided by Pittendrigh's inference that the resetting map is instantaneously shifted in phase by light pulses 2 or 3 orders of magnitude stronger than those used here. Since this model [**1**] involves a 1-variable (phase) description of the state of the clock, it seemed a plausible conjecture that the latter is strongly self-exciting: a quickly-recovering limit-cycle or a relaxation oscillator. This is in fact the basis of the formally identical entrainment model independently derived by Perkel, et al. for pacemaker neurons [**11**], [**12**].

However the plausibility of this hypothesis is reduced by the ready accessibility and apparent stability of the singularity and by the new fact that the shape of the resetting map can be distorted by prior non-phase-shifting perturbations. The accuracy of the classical 1-variable description may derive less from the character of unperturbed clock dynamics than from a peculiarity of the effect of very strong perturbations, such as is incorporated in the model of Pavlidis [**8**].

18. Circadian clocks in other organisms

Is the general picture sketched thus far unique to Drosophila, or might it be pertinent to circadian processes in other organisms as well? The experimentally determined surface, $\theta(T,S)$, may be regarded as a 1-parameter family of resetting maps, where S, the duration of the perturbation, is the selecting parameter. The peculiar shape of this surface—a spiral ramp connecting tilted lamina—makes this a very definitely constrained family of shapes. We already know that it contains only Type 0 and Type 1 curves; moreover, those curves have particular shapes. We choose two simple measures of shape: the peak-to-peak amplitude of the resetting map, PP, and the range of T between the peaks, W. For Type 0 resetting maps, we measure from a horizontal baseline, and for Type 1, from a 45° baseline. For Type 0 resetting maps we will measure W across the downslope between peaks, and for Type 1, across the upslope (Figure 15).

It can be shown that for smooth, simple surfaces of the sort shown in Figure 3, PP varies smoothly from 0 to $\frac{1}{2}$ (*not* to 1) and W varies from $\frac{1}{2}$ to 0 as S increases from 0 to

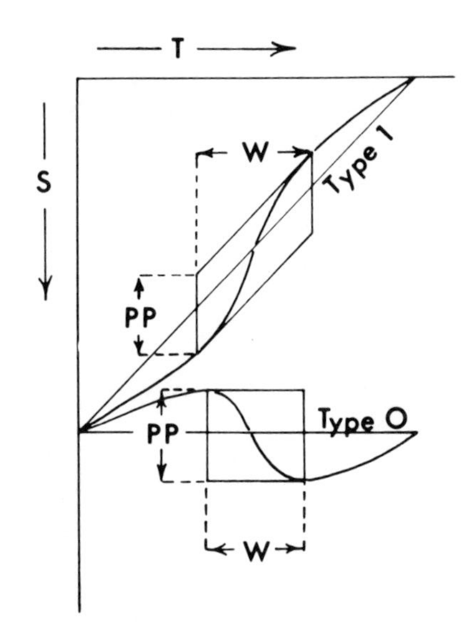

FIGURE 15
Peak-to-peak amplitude (PP) and width (W) measurements, applied to Type 1 and to Type 0 resetting maps. Note that W is not *defined* as the smaller of the two possible W measurements, though it always turns out to be the smaller.

S^*. Then within Type 0, the same changes take place in reverse, terminating at large S with PP $= 0$ and $W = \frac{1}{2}$.

Figure 16 provides polar coordinates for a plot of PP vs. W loci, with Type 1 curves on the left and Type 0 to the right of the vertical centerline. PP increases radially from the center of the larger circle, of radius $\frac{1}{2}$ ($= 12$ hours). The angular coordinate is W, increasing on each side from 0 at the bottom to 1 at the top. The smaller circle is a simple representation of the kind of PP vs. W locus characteristic of cophase surfaces resembling Figure 3, as discussed above.

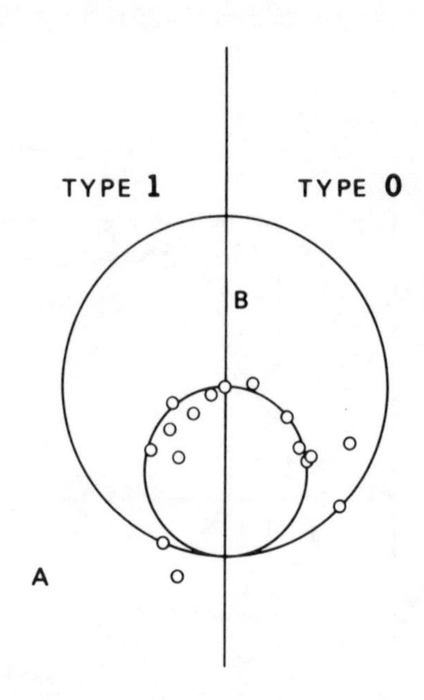

FIGURE 16
A polar coordinate plane (described in text) for plotting resetting map shape measurements, (PP, W). The larger circle marks PP $= \frac{1}{2}$; the smaller is the approximate (PP, W) locus anticipated if clock dynamics resembles Figures 7 and 8.

The tiny circles represent the shapes of the 16 "best" (Table 1) resetting maps available in the literature. Shapes far from the idealized locus are conspicuously absent, even though such shapes may seem perfectly plausible a priori. For example A in Figure 16 would be a sinusoid of amplitude ± 9 hours with the extremes 6 hours apart on the T axis. B would be a sawtooth with a near-discontinuity of 6 hrs. Neither has occurred as far as I know.

In three pairs of cases, resetting data are given for the same organism and kind of perturbation, but at two durations; in all cases the shape changes in the anticipated way (counterclockwise) around this locus as the duration is increased.

It may be therefore that the circadian oscillators which regulate diverse physiological and behavioral functions in many phyla of plants and animals, as well as unicellular organisms, all derive from smooth and simple rotatory flow around a singularity in a state space of at least two

TABLE 1

Type 1 Resetting Maps			Type 0 Resetting Maps		
PP (hours)	W (hours)	Ref.	PP (hours)	W (hours)	Ref.
0	—	31	8	7	26
1	9	20	12	6	27
3	7	21	8	7	28
5	7	22	7	7	29
7	7	23	10	9	1
6	5	23	5	9	9
10	4	24	4	13	30
12	3	25	0	—	—[6]
13	2	26			

[6] Very short perturbation has no effect.

degrees of freedom, and are affected by light in qualitatively similar fashions.

19. (T^*,S^*) as an experimental tool

There may be some value, beyond mere demonstration of its existence, in possession of a recipe for sending an oscillatory process of unknown and nonlinear complexity to its stationary state, especially if that state is stable in the sense that trajectories through nearby states do not rapidly leave the neighborhood.

(a) If the critical perturbation is brief, as in the case illustrated, then one can immediately inhibit all change in the state variables of the basic oscillator by a brief alteration in its environment. Any quantity subsequently found to change is therefore excluded as a potential component of the basic clock process or even as a factor affecting that process. (This argument could be invalidated by demonstrating a third important degree of freedom.)

(b) It becomes possible to examine the behavior of clock-influenced processes (e.g., termination of metamorphosis) in the absence of the driving or synchronizing clock oscillations, but without permanently changing the environment.

(c) Factors which influence clock dynamics only very weakly (e.g., most metabolic inhibitors, unless used in pathologically large doses) cause only small phase shifts in pulse-chase experiments. However, the same small displacement of state applied when the oscillator is at the singularity would initiate a distinct rhythm. Moreover, all factors impinging on the clock in the same way at the singularity will initiate the rhythmicity at the same phase. The number and nature of equivalent factors may suggest the number and nature of the discrete processes or variables affected.

Summary

A simple and general method for exploring the dynamics of fixed-period oscillatory processes is surprisingly successful in predicting unlikely, but testable, behavior of the circadian clock of the fruitfly, Drosophila pseudoobscura. This clock regulates the time of emergence of the adult fly from the pupal case; this time can be advanced or delayed by a light pulse, although after transients the period of the clock remains the same (about 24 hrs). The size of the phase-shift depends upon *when* in the clock's cycle we give the light perturbation. A resetting map plots the time of emergence (cophase) against the time (T) when a perturbation of duration S is given.

We construct the most general, simple oscillator consistent with two empirical generalizations derived from published data. This "oscillator" is a description of dynamical flow in an abstract state space. By analysis of its response to pulse perturbations, it was inferred that:

(1) As the length (S) of the light pulse increases, the resetting map will change discontinuously from Type 1 (average slope over a cycle $= 1$) to Type 0 (average slope $= 0$), and no other types of resetting maps will be found. Furthermore, as S increases, the shape of the Type 1 resetting maps (a straight 45° line for $S = 0$) will pass through a continuous but restricted family of shapes until the switch to Type 0 resetting maps, which gradually

smooth out to a straight horizontal line for large S. (There *could* be a reversion to Type 1 if $N \geqq 3$.)

A three-dimensional graph of cophase as a function of T and S will resemble a stack of tilted planes connected by a right-handed helicoid. All possible resetting maps will be sections at constant S through this graph.

(2) There exists a critical annihilating light pulse, (T^*, S^*), which places the clock in a "phaseless", non-oscillatory state (perhaps a singularity in the dynamical flow), although *longer* pulses merely phase-shift the clock without permanently affecting its normal behavior. This critical pulse is the symmetry axis of the above helicoid.

(3) This critical pulse does not destroy the clock, for if a second pulse (S) follows, the clock will resume oscillations as if a *single*, phase-shifting (T^*, $S^* + S$) pulse had perturbed it. The effect of this second pulse depends entirely on its length; *when* it is given is irrelevant, unless $N \geq 3$.

Experiments have been completed which appear to confirm these expectations for $N = 2$ in the Drosophila pseudoobscura clock.

It is deduced from these data that the state of the clock can vary simultaneously in at least 2 independent ways and it may have *only* 2 important independent variables, since the phaseless state appears to be time-independent, i.e., a true singular state.

Moreover, since the singularity appears stable, and indeed, the pupal clocks in the dark seem to be created "at rest", clock models based on relaxation oscillators or very stable, self-exciting limit-cycle oscillators, for example, may no longer be relevant for Drosophila.

The fact that the resetting maps of other organisms have similar shapes and can be grouped as Type 1 or

Type 0, Type 0 corresponding to longer perturbations, leads us to speculate that this kind of analysis of perturbation experiments may be useful for work on the oscillatory processes of other organisms.

The possible role of the singularity as a tool for further experiments is also discussed.

Acknowledgements

I am grateful to Jay Mittenthal, who introduced me to the idea of a dynamical space, and to the many workers in the "clocks" field who shared with me their response-curve data and experimental techniques. In particular I thank Colin Pittendrigh, who made it possible for me to pursue these experiments in his lab. The flies were constructed from mutant chromosomes kindly provided by Ronald Quinn and Theodosius Dobzhansky. I am indebted to Theodosios Pavlidis for stimulating discussion, and through his essays, "Strong Inference" and "The Art of Creative Thinking", to John Platt for the inspiration of this whole project.

This work was done in partial fulfillment of the requirements of the Ph.D. in biology at Princeton University while enjoying the support of an NSF Regular Fellowship. The laboratory was supported under NONr-1858(28) and NASr-223.

REFERENCES

1. C. S. Pittendrigh, *Circadian clocks*, edited by: J. Aschoff, North-Holland, Amsterdam, 1965, p. 277.

2. T. Pavlidis, Bull. Math. Biophys. **29** (1967), 291.

3. E. Ottesen, Bach. Thesis, Biology Dept., Princeton University, Princeton, N.J., 1965.

4. A. T. Winfree, Ph.D. Thesis, Biology Dept., Princeton University, Princeton, N.J., 1970.

5. ———, J. Theor. Biology **16** (1967), 15.

6. ———, Proc. Sympos. (Prague) edited by: E. K. Pye (to appear).
7. C. S. Pittendrigh, Proc. Nat. Acad. Sci. U.S.A. **58** (1967), 1762; 1862.
8. T. Pavlidis, *Studies on biological clocks: A model for the Circadian rhythms of nocturnal organisms*, Lectures in Math. Life Sciences, vol. 1, Amer. Math. Soc., Providence, R.I., 1968, pp. 88–112.
9. H. W. Honegger, Zeit. für Vergl. **57** (1967), 244.
10. W. Zimmerman, Biol. Bull., **136** (1969), 494.
11. D. H. Perkel, et al, Science **145** (1964), 61.
12. R. Fitzhugh, Biophys. J. **1** (1961), 445.
13. A. Kalmus and L. A. Wigglesworth, Cold Spring Harbor Sympos. on Quant. Biology **25** (1960), 211.
14. J. R. Platt, Science 146 (1966), 347.
15. ———, *The excitement of science*, Houghton-Miflin, Boston, Mass., 1962.
16. C. S. Pittendrigh, Zeit. für Pflanz. **54** (1966), 275.
17. A. T. Winfree, Unpublished experiments.
18. A. Campbell, *Synchrony in cell division and growth*, edited by: E. Zeuthen, Interscience, New York, 1964, p. 469.
19. W. Engelmann and H. W. Honegger, Zeit. für Naturfor. **22b** (1967), 200.
20. P. DeCoursey, Science **131** (1960), 33.
21. J. Feldman, Ph.D. Thesis, Biology Dept., Princeton University, Princeton, N.J., 1967.
22. D. Minis, *Circadian clocks*, edited by: J. Aschoff, North-Holland, Amsterdam, 1965, p. 333.
23. J. Burchard, Ph.D. Thesis, Biology Dept., Princeton University, Princeton, N.J., 1958.
24. J. Feldman, Proc. Nat. Acad. Sci. U.S.A. **57** (1967), 1080.
25. M. Sargent and W. Briggs, Proc. Nat. Acad. Sci. U.S.A. **58** (1967), 1862.
26. R. Halaban, Plant Physiol. **43** (1968), 1887.
27. J. Hastings, *Photophysiology*, edited by: A. Giese, Academic Press, New York, 1964, p. 333.
28. D. Horne, Bach. Thesis, Biology Dept., Princeton University, Princeton, N.J., 1967.
29. R. Zimmer. Planta **58** (1962), 283.
30. V. Bruce, F. Weight and C. S. Pittendrigh, Science **131** (1960), 728.

AUTHOR INDEX

Roman numbers refer to pages on which a reference is made to a work of an author.

Italic numbers refer to pages on which a complete reference to a work by an author is given.

Boldface numbers indicate the first page of an article in this volume.

SUBJECT INDEX